# Creating Diversity Capital

# Creating Diversity Capital

*Transnational Migrants in Montreal,
Washington, and Kyiv*

Blair A. Ruble

Woodrow Wilson Center Press
Washington, D.C.

The Johns Hopkins University Press
Baltimore

EDITORIAL OFFICES
Woodrow Wilson Center Press
Woodrow Wilson International Center for Scholars
One Woodrow Wilson Plaza
1300 Pennsylvania Avenue, N.W.
Washington, D.C. 20004-3027
Telephone: 202-691-4010
www.wilsoncenter.org/press


ORDER FROM

The Johns Hopkins University Press
Hampden Station
P.O. Box 50370
Baltimore, Maryland 21211
Telephone: 1-800-537-5487
www.press.jhu.edu/books/

2 4 6 8 9 7 5 3 1

Library of Congress Cataloging-in-Publication Data

Ruble, Blair A., 1949–
   Creating diversity capital : transnational migrants in Montreal, Washington, and
Kyiv / Blair A. Ruble.
      p.   cm.
   Includes index.
   ISBN-13: 978-0-8018-8300-8 (hardcover : alk. paper)
   ISBN-13: 978-0-8018-8301-6 (pbk. : alk. paper)
      1. City dwellers.   2. Emigration and immigration—Social aspects—Case studies.
3. Pluralism (Social sciences)   4. Hybridity (Social sciences)   5. Social change.
6. Montréal (Québec)—Emigration and immigration—Social aspects.   7. Washington
(D.C.)—Emigration and immigration—Social aspects.   8. Kiev (Ukraine)—Emigration
and immigration—Social aspects.   I. Title.
HT215.R83   2005
307.76′092—dc22
                                                                              2005022601

**Woodrow Wilson
International
Center
for Scholars**

The Woodrow Wilson International Center for Scholars, established by Congress in 1968 and headquartered in Washington, D.C., is a living national memorial to President Wilson. The Center's mission is to commemorate the ideals and concerns of Woodrow Wilson by providing a link between the worlds of ideas and policy, while fostering research, study, discussion, and collaboration among a broad spectrum of individuals concerned with policy and scholarship in national and international affairs.

Supported by public and private funds, the Center is a nonpartisan institution that establishes and maintains a neutral forum for free, open, and informed dialogue. Conclusions or opinions expressed in Center publications and programs are those of the authors and speakers and do not necessarily reflect the views of the Center staff, fellows, trustees, advisory groups, or any individuals or organizations that provide financial support to the Center.

The Center is the publisher of *The Wilson Quarterly* and home of Woodrow Wilson Center Press, *dialogue* radio and television, and the monthly newsletter "Centerpoint." For more information about the Center's activities and publications, please visit us on the Web at www. wilsoncenter.org.

*Pro urbis amore*

# Contents

# Tables

# Acknowledgments

Unscrew the locks from the doors!
Unscrew the doors themselves from their jambs!

—Walt Whitman, *Song of Myself,* 1855 edition

This volume was inspired by the scholarship, life, and memory of Galina Vasil'evna Starovoitova, a member of the Russian State Duma who was murdered on the staircase leading to her Saint Petersburg apartment on the evening November 20, 1998. Galina profoundly believed in the fundamental dignity of human existence. She was as forceful a defender of individual liberty as Russia has produced, and she was a friend as well as a colleague.

Galina and I both spent the better part of the 1980s separately conducting research and writing about her home city, which was then known as Leningrad. We followed and read each other's work closely, although the realities of Soviet security paranoia kept us from meeting one another in person until the end of the decade.

Galina's book, *Ethnic Groups in a Contemporary Soviet City,* inspired my own research on the diverse character of Russian cities over subsequent years. Much to my good fortune, Galina came to spend a month at the Kennan Institute in Washington as a Woodrow Wilson Center Guest Scholar shortly after I had become the Institute's director. I feel honored to have had the opportunity to work with her.

For those who knew Galina—even if only through televised parliamentary debates—she was a force of nature. She once confided in me following an appearance at the Council on Foreign Relations in New York City that Henry Kissinger had been in the front row. She proudly added that Dr. Kissinger had called her "a real tub thumper." Walt Whitman's admonition to throw open the doors is in keeping with Galina's spirit.

For me, Galina will always be the urban specialist whose academic work inspired even as she emerged as a prominent public figure of considerable consequence in Russia. Whenever given the opportunity, she would encourage me to explore how members of ethnic communities transform the cities in which they live—for the better.

I would never have been able to contemplate—let alone write—this volume had it not been for the interest of others. I therefore would like to acknowledge the encouragement, and intellectual and research support, for this project by Kennan Institute interns Valerie J. Chittenden, Sapna Desai, Oksana Klymovych, Jarom McDonald, Janet M. Mikhlin, Mary Frances Muzzi, Cynthia Neil, Olena Nikolayenko, Luba Shara, and Oliya S. Zamaray. I also would like to express my gratitude to Kengo Akizuki, Dominique Arel, Olena Braychevska, David Biette, Caroline Brettell, Joseph Brinley, Jennifer Giglio, Howard Gillette Jr., Lisa Hanley, Stephen E. Hanson, Olena Malynovska, Heather McClure, Oleksandr Mosjyk, Oxana Shevel', Margaret Paxson, Mario Polese, Nancy Popson, Yaroslav Pylynsky, Brian Ray, Richard Stren, Joseph S. Tulchin, Diana Varat, Oleksandr Vazylenko, and Galina Volosiuk for their thoughtful comments, suggestions, and criticisms during the writing of this book. My editors with the Woodrow Wilson Center Press, Yamile Kahn and Alfred F. Imhoff, have saved this volume from my infelicitous prose on several occasions, more than earning my heartfelt gratitude.

Dominique Arel, probably the leading Québécois Ukrainianist to be found anywhere, has been especially supportive of my efforts to write about two beloved cities that he knows as well as anyone. Dominique has encouraged when necessary, criticized when necessary, and inspired when necessary. He has proven to be a trusted colleague and close friend. The pages to follow reveal just a portion of the tremendous intellectual, professional, and personal debts that I owe him.

David Biette, my colleague who serves as director of the Woodrow Wilson Center's Canada Institute, as well as Heather McClure and her colleagues at Washington's Council of Latino Agencies were especially helpful to me during my work on the Montreal and Washington portions of the research for the book. I alone remain responsible for the (alas) inevitable miscues that somehow have made it past Dominique, David, Heather, and a brood of other beleaguered reviewers.

Although we have never met, I would like to thank La Bloggeuse Volubile at the Montreal City Web Log (http://w5.montreal.com/mtlweblog). Cities everywhere would be better places if every metropolitan community were so fortunate as to have such a rich informational resource sharing their news with the world at large.

I would like to pay special homage to my partners in the organization and implementation of the Kennan Kyiv Project's surveys of transnational migrants, native residents, and migration specialists living in Kyiv. Working with Olena Braichevska, Olena Malynovska, Nancy Popson, Yaroslav Pylynskyi, and Galina Volosiuk on surveys related to migrant issues in Kyiv was a distinct honor. Our collaborative research provides much of the data upon which the sections on the Ukrainian capital have been based. Further discussion of the conduct of those surveys may be found in the appendix to this volume.

These surveys were supported by the George F. Kennan Fund of the Kennan Institute of the Woodrow Wilson International Center for Scholars, with the assistance of the U.S.–Ukraine Foundation and the Office of the United Nations High Commissioner for Refugees (UNHCR) in Ukraine. I am deeply grateful to the participating transnational migrant communities in the city of Kyiv, especially activists from their community associations and interpreters, as well as to the representatives of the U.S.–Ukraine Foundation and UNHCR for their kind assistance throughout the Kyiv portion of this work.

The Kennan surveys followed on a moment of social science serendipity that has enriched this project from the very beginning. In 1998 and 1999, Nancy Popson—then the Kennan Institute's deputy director—was spending as much time as she could visiting various neighborhood elementary schools around Kyiv to see how history was being taught in an independent Ukraine. Nancy returned to the Kennan Kyiv Project Office full of excitement following a day of visiting classrooms in Troeshchyna. She had spent the afternoon at School 247 with children from Afghanistan, Angola, Mongolia, and Vietnam.

Nancy's encounter with School 247 prompted both of us to begin to inquire about these children, only to discover a growing community of transnational migrants settling in Troeshchyna and elsewhere around the city. To our great good fortune, Nancy and I—together with Kennan Kyiv Project Director Yaroslav Pylynsky—formed a research alliance with three remarkable Ukrainian specialists on migration, Olena Braychevska, Olena Malynovska, and Galina Volosiuk.

The results of our collaboration appear on the pages about Kyiv to follow, as well as in several previously published English language works, including Nancy E. Popson and Blair A. Ruble, "Kyiv's Nontraditional Immigrants," *Post-Soviet Geography and Economics* 41, no. 5 (2001): 365–78; Nancy E. Popson and Blair A. Ruble, "A Test of Urban Social Sustainability: Societal Responses to Kyiv's 'Non-Traditional' Migrants," *Urban Anthropology* 30, no. 4 (2001): 381–409; Blair A. Ruble, "Kyiv's Troeshchyna: An Emerging International Migrant Neighborhood," *Nationalities Papers* 31, no. 2 (2003): 139–55; and Olena Braichevska, Halyna Volosiuk, Olena Malynovska, Yaroslav Pylynskyi, Nancy Popson, and Blair Ruble, *Nontraditional Immigrants in Kyiv* (Washington, D.C.: Woodrow Wilson Center Comparative Urban Studies Project and Kennan Institute, 2004). More complete findings are to be found in a Ukrainian language volume: O. Braichevska, G. Volosiuk, O. Malynovska, Ia. Pylynskyi, N. Popson, and B. Ruble, *Netraditsiini immigranti u Kyivi* [*Non-Traditional Immigrants in Kyiv*] (Kyiv: Kennan Kyiv Project, 2003).

I became increasingly drawn to the example of Montreal as I tried to consider the findings emerging from our surveys in Kyiv. The choice was a natural one, given the Canadian city's intricate dance of language and politics with growing transnational migrant communities. Beyond the mere presence of migrants, Montreal proved to be an ideal foil to Kyiv because of an extensive and impressive social science literature examining the city that has appeared during the past four decades. This work frequently explored issues that are fundamental to understanding the emerging situation in the Ukrainian capital.

"Life itself," as a certain type of Soviet-era writer might put it, offered up the Washington example for my consideration. Living and working in the District of Columbia, I knew the city to be an urban community far more complex and rich in history and nuance than its superficial image as the U.S. capital might suggest. A brief portion of this text appeared as an "op-ed" article in a Washington newspaper: Blair Ruble, "From 'Chocolate City, Vanilla Suburbs' to 'Rocky Road,'" *Washington Afro-American*, October 16–22, 2004.

The result of these odd, somewhat chance connections appears on the pages that follow. My goal has been to use these examples as a provocation. I want to prompt my readers to think about cities and transnational migrants a little differently after finishing this book than they did when they picked it up for the first time.

If I have succeeded in this goal, I have done so only because of all the people mentioned in these inadequate acknowledgments, and many others as well. The fact that I was able to conceive of this book in the first place pays tribute to my years of friendship with Galina Starovoitova. This modest volume represents my small effort to continue her academic legacy, and to honor her memory.

Washington, D.C.
June 1, 2005

# Part I

# Introduction

# Chapter 1

# Creating Diversity Capital:
# Migrants in Divided Cities

The original settlers knew that the only way human beings could live together here was by practicing tolerance. Sometimes it was a reluctant tolerance. Sometimes it was tolerance combined with the wink and shrug of hypocrisy. But in the end, it was a tolerance that insisted on one fundamental truth: There are people here who are not like us, and we must accept them in order to live.

—Peter Hamill, *Downtown: My Manhattan,* 2004[1]

The New York University historian Thomas Bender, in a collection of essays commemorating the first anniversary of the September 11, 2001, terrorist attack on the World Trade Center, explored the meaning of his city within an American—and, by extension, global—context. "The special character of New York was evident from the beginning," Bender observed. "If religion inspired the Puritans," he continued, "and if the dream of plantations and wealth drove the Virginians, the practicality of trade engaged the first settlers of New Amsterdam. If churches and regular church service came quickly to both Massachusetts and Virginia, it was the countinghouse, not the church, that represented early New Amsterdam. In fact, the first substantial building in Manhattan was a stone countinghouse. There was little impulse to exclusion; trading partners were sought no matter what their background. Already in the 1640s eighteen languages were spoken in the area that is now New York City."[2] "This very different history," Bender goes on to argue, "became the material for a different understanding of society and politics, one that embraced difference, diversity, and conflict—as well as the dollar. The city was characterized by a divided elite and a rich diversity of groups and cultures. As a result, the city early experienced a

3

continuing contest over the definition of itself."[3] For Bender, New York's search for itself is continuing in the wake of the tragic events of September 11. Like the nineteenth-century New York poet Walt Whitman, Bender believes that the city's meaning can best be found in its grand cacophony.

At a most superficial level, the origins of New York's embrace of difference originates at New Amsterdam's early stone countinghouse and the eighteen languages swirling around it. Russell Shorto, in an homage to the Dutch legacy of tolerance and democracy that he sees as undergirding all that New York would become, emphasized the settlement's startling diversity from the very outset. Shorto writes:

> It was founded by the Dutch, who called it New Netherland, but half of its residents were from elsewhere. Its capital was a tiny collection of rough buildings perched on the edge of a limitless wilderness, but its muddy lanes and waterfront were prowled by a Babel of peoples—Norwegians, Germans, Italians, Jews, Africans (slaves and free), Walloons, Bohemians, Munsees, Montauks, Mohawks, and many others—all living on the rim of empire, struggling to find a way of being together, searching for a balance between chaos and order, liberty and oppression. Pirates, prostitutes, smugglers, and business sharks held sway in it. It was *Manhattan,* in other words, right from the start: a place unlike any other, either in the North American colonies or anywhere else.[4]

New Amsterdam's Tower of Babel in miniature was symptomatic of deeper structural properties. The city's commercial elite became too divided at the very outset of its existence to establish secure boundaries between those who were to be included in their community and those who were not. Bender notes that "you had a place in a [New England] Puritan village or town only if your values coincided with those of your neighbors. Rather than incorporating difference, Puritan town leaders were quick to offer strangers the 'liberty to keep away from us.'"[5] New Amsterdam's corporate overseers dispatched by the Dutch West India Company had scant choice but to welcome all nature of beings to their mean and inconsequential settlement if they were to meet the demands for profit emanating from shareholders back in the Netherlands. Pragmatism, rather than community or virtue, became the order of the day.

As different groups arrived in New Amsterdam to seek fame and fortune, power became divided, dispersed, and contested (witness the growing competition between Dutch overseers and rival English settlers from New England who increasingly dominated the colony's economy). Before too long, even a chief executive as tyrannically authoritarian as Governor Peter Stuyvesant could no longer command compliance to his will. In fact, no single social, political, economic, or ethnic group exhibited a capacity to exert secure domain over others.

Stuyvesant's great rival Adriaen van der Donck eventually convinced the colony's trustees back in Holland to grant a measure of home rule. The consequent town charter established a local government of sorts in 1653, with power becoming grudgingly shared by the Dutch West India Company and local merchants. As Shorto tells the tale, "With a rudimentary representative government in place, the island rapidly came into its own. Stuyvestant and the West India Company still officially ran the place, but, whether they were Dutch, English, or any of the other nationalities represented in the colony, the businessmen—the fur traders, the tobacco farmers, the shippers of French wines, Delft tiles, salt, horses, dyewood, and a hundred other products—increasingly got their way."[6]

Local politics became neither a means to pursue virtue nor to sustain community. Managing New Amsterdam—and the English colonial town to follow—required an at times forced accommodation of private interest. "What matters about the Dutch colony," Shorto tellingly observes, "is that it set Manhattan on course as a place of openness and free trade."[7] This continued to be the case following the arrival of British rule in 1664. The wily Stuyvestant negotiated recognition by the town's new English occupiers of the same rights and privileges to self-government that he, as local director of the Dutch West India Company's assets, had so vigorously opposed throughout his embittered battles with local burghers.[8] To achieve any set of personal or group goals meant transcending a "zero-sum game" by engaging with others to pursue shared objectives.

Such an embrace of "pragmatic pluralism" did not necessarily foster democratic institutions and goodwill among human beings.[9] The practice of pragmatic pluralism ignores questions of virtue and community. A community, by its very nature, is an exclusive form of social organization that is notoriously intolerant of those not already among the embraced. Pragmatic pluralism emerges from a willingness born of necessity to tolerate behavior that is to some degree offensive. Pragmatic pluralism—and the survival strategies associated with it that are so visible in as robustly cosmopolitan a city as New York—glorifies the middle ground that exists among all residents and communities. And by doing so, it expands a city's capacity to adapt to diversity.

## Creating Diversity Capital

Commercial entrepôts such as Amsterdam, and its offspring New York, are predicated on the tumbling together of diverse populations for private gain.[10] They are marked almost from birth by high levels of a characteristic that Richard Stren of Toronto and Mario Polese of Montreal have called "urban social sustainability." For Stren and Polese, urban social sustainability consists of "policies and institutions that have the overall effect of integrating diverse groups and cultural practices in a just and equitable fashion."[11] Those communities that achieve higher levels of urban social sustainability are those that are able to expand their repertoire of equitable responses to diversity.

Urban meeting places such as an Amsterdam or a New York have demonstrated a high capacity to accommodate diversity for decades, even centuries. Yet their histories are not always seamless. Stren and Polese's standard of "just and equitable" often represented an objective beyond the grasp of local residents. Indeed, Amsterdam's politics has taken a distinctly intolerant turn as the twenty-first century begins. Nonetheless, both cities have proven themselves on balance to be organic venues for the forced accommodation of difference through the practice of pragmatic pluralism. In comparison with much of the world, Amsterdam and New York are communities with a deep-seated capacity for absorbing diversity that is rooted in their long evolution over centuries. They are cities, in other words, with what might be considered to be high stocks of "diversity capital."

A diversity capital perspective inverts notions of social, cultural, and financial capital that were developed by Pierre Bourdieu during the 1970s and 1980s (and that since have been expounded upon by numerous social scientists).[12] Bourdieu resurrected notions of social capital to help explain the maintenance and reproduction of the power and influence of dominant groups over society. He presented various forms of capital as interchangeable, with social and cultural relationships being used to obscure the transfer of wealth and influence among dominant groups.

Other social scientists—most notably Robert Putnam—have used the concept of social capital to explore how individuals and groups mobilize social resources to pursue a variety of cultural, economic, and political goals.[13] The concept has become remarkably murky, with debates erupting over whether or not there can be both "good" and "bad" social capital: social relationships built on trust in the state that reinforce a virtuous cycle enhancing state capacity versus ones built on trust in opposition to the state that consequently undermine official authority.[14]

Following on the very different work of writers such as Bourdieu and Putnam, an academic mini-industry has emerged seeking to determine the extent and quality of social capital shared among various groups and individuals. This approach assumes a larger system (social, elite, state) with which the groups and individuals interact. The resulting studies provide a "bottom-up" view of how communities and individuals mediate their own interests with those of a larger socioeconomic system or state. When considering migrants, for example, the analysis of social capital often focuses on their integration or nonintegration into host societies.[15]

By contrast, this volume explores the impact of transnational migrant communities on the host societies of three cities: Montreal, Washington, and Kyiv. In general terms, it argues that the presence of these communities—and an accompanying creation of what is being identified as "diversity capital"—disrupts long-standing systems of metropolitan dominance. As a consequence, the focus falls on how the presence of these migrant communities generates an openness toward out-groups ("diversity capital") rather than reinforcing in-group forms of "trust" ("social capital").

Although this study acknowledges that the literature on social capital is richer and more textured than what is presented here, it nonetheless attempts to isolate broad social processes in a different way. By examining the recent histories of Montreal, Washington, and Kyiv, it seeks to explore how the arrival of migrant communities has transformed urban social, economic, and governance systems rather than strengthened existing power relationships.

This perspective shifts the focus of attention from individual migrants or migrant groups to urban communities of metropolitan scale. By doing so, this effort to better explain the place of migrants and migrant communities in metropolitan regions implicitly turns the social capital literature to date on its head. Migrants reshape the city rather than the city assimilating migrants (or not).

The quality identified here as "diversity capital" is an especially valuable resource in the early years of the twenty-first century. The present is a moment of rapid and large-scale movement of people around the globe.[16] Migration, both "regular" (legal) and "irregular" (illegal), has become a major concern in "receiving" and "sending" countries alike throughout both the advanced industrialized and nonindustrialized worlds.[17] Diasporic cultures—often formed by the embittering and frustrating experience of displacement and exile—are powerful forces shaping the societies and cultures of both the countries left behind and those now home to twenty-first-century "global villagers."[18] Not surprisingly, transnational migrants are transforming almost every kind of community that can be said to exist in the world. They continue to do so even in the wake of the profound changes in the international system that have followed from the events of September 11, 2001.

Transnational migrants have come to represent new challenges for nearly all the world's great cities because it is precisely in cities—and especially in large cities and metropolitan regions—where the existence of a plurality of interests, identities, communities, and individuals can no longer be denied.[19] Traditional conceptions of place and community are collapsing under the uncommon challenges of rapid metropolitan urban growth, instantaneous communication, and rapidly moving people.

Sustaining a civic consciousness beyond group identity in an age preoccupied with speed and velocity has become no humble task. A new era of metropolitan diversity disrupts previous understandings of power. Social groups in today's cities are forced to choose their ground carefully, moving to protect interests only in those areas that are understood to matter for their survival or well-being. Municipal life thus becomes an at times forced accommodation of competing private interests.

Martin Heisler has argued that migration represents an especially advantageous point of departure for studying societal transformation precisely because of its mass scale and its potent influence over all people and institutions at every level of society.[20] Heisler contends further that migrants—foreigners, strangers, outsiders—become convenient markers for change. Migrants, he continues, disturb the status quo in multiple, unpredictable, and powerful ways, thereby both exposing what a society thinks it once was and provoking disputes over what a society should become in the future.

How do urban communities adjust as they accommodate the new realities of this century's massive transnational migrations? At first blush, the answer to the twenty-first-century conundrum of social diversity would appear to be found in entrepôts such as Amsterdam and New York, urban communities that have accumulated diversity capital over their long histories. Such an approach, though illuminating, runs the danger of tautology: New York has accommodated a rich diversity of residents because it has done so since its founding in the early seventeenth century. Such observations fall short of revealing how diversity capital might be created in urban communities with less tolerant histories, even as they may help to explain why an Amsterdam or a New York may be well situated for success in the present historical moment.

Therefore, this volume seeks clues about how a city's capacity for urban social sustainability may expand by examining more classically segregated communities that have been rapidly accumulating new diversity capital during recent decades. In other words, it tries to discern moments when the qualities so evident in historically tolerant cities such as Amsterdam and New York—recognizing all the limitations to that tolerance at various moments in the past as in the present—emerge in even more hostile urban environments. This study does so by examining the impact on communities long divided by language and race of unprecedented numbers of arriving migrants from diverse ethnic, racial, linguistic, confessional, cultural, and economic backgrounds. It explores moments in the recent histories of Montreal, Washington, and Kyiv when the arrival of more and more varied residents has expanded their overall capacity to accommodate diversity.

More particularly, this book tells the story of the presence of tens of thousands of transnational migrants in Montreal, Washington, and Kyiv as a means for exploring the ways in which new diversity capital can be accumulated and used (see note for further discussion of the term *transnational migrants*).[21] If successful, the following pages will convince the reader that many cities—beyond traditional cosmopolitan centers such as Amsterdam or New York—have rich possibilities for accommodating diversity.

Montreal, Washington, and Kyiv are ideal sites for examining the creation of diversity capital. Language framed the cultural, social, and political landscapes of both Montreal and Kyiv, as competition and conflict raged between French and English, Ukrainian and Russian. Their urban realities become defined through mutually exclusive linguistic categories, which are thought to supersede neighborhood, occupation, gender, and class. Similarly, Washington has been preoccupied by race ever since the fledgling U.S. government claimed the plantations of slaveholding Virginians and Marylanders for its new capital. Despite these differing pasts, all three cities have experienced dramatic and rapid recent increases in the number of transnational migrants moving in to live and work. Therefore, they provide an opportunity to consider how diversity capital is created at a moment when the process is under way.

## Montreal: From "Two Solitudes" to "Many Solitudes"

The Canadian novelist Hugh MacLennan defined the reality of French and English Canada in his landmark 1945 novel *Two Solitudes* as parallel universes living in uneasy tension with one another along the shores of the Saint Lawrence River Valley. "But down in the angle at Montreal," he begins his tale of isolation, "two old races and religions meet here and live their separate legends, side by side."[22]

The sense of Montreal as being divided between two parallel yet distinct solitudes, of "two old races and religions" that meet and live "their separate legends, side by side," dates back to the French abandonment/English conquest of Quebec in 1763. Montreal had already fallen under British control in 1760, a year after the British general James Wolfe's defeat of French forces under Louis Montcalm on the Plains of Abraham just beyond the walls of the French citadel at Quebec City. The British were far more intent on securing durable control over the region than had been the French, who relied for sustenance on open-ended trading relations with the indigenous Iroquois and other native peoples of the region. Anglo-Protestant and Franco-Catholic have uneasily cohabited in Montreal ever since.

Like race in the United States, language—and religion, which often lies behind the language issue—constitutes a founding uncertainty to which Canada has yet to find a seemingly permanent accommodation. Indeed, language, culture, and religion have combined with geography and history to produce at times unbridgeable chasms running through Canadian society. As the political scientist Rejean Landry noted during the late 1970s, Canadian history is, to a considerable degree, the product of constant intergroup negotiation between the French and the English.[23]

As Canada's largest city for some two centuries—before falling behind Toronto during the early 1970s[24]—and as one inhabited by both linguistic groups, Montreal has remained a central venue for Canada's unending renegotiation with itself.[25] The political scientist Harold Kaplan highlighted this essential aspect of Montreal's life in his monumental study of Canadian urban political culture—*Reform, Planning and City Politics: Montreal, Winnipeg, Toronto*—when he noted that Montreal "was the cockpit of French/English relations in Canada and the inheritor of both long-standing and more recent animosities."[26] Kaplan continued that "Montreal city politics was of interest not in its own right, but as a subordinate arena, a stage on which larger national questions would be played out."[27] The result was an "intense, warlike politics" that "pervades all parts of the community. There are [in Montreal] friends and enemies, but nothing in between."[28]

The complex history of language, religion, and ethnicity in Montreal rests to a considerable degree on the postconquest British decision to permit the French residents of Upper Canada to retain their core institutions: the Roman Catholic Church, schools, and civil code. French Canadian life came to be centered on small parishes and rural settlements, leaving Montreal to colonial administrators. Montreal became a predominantly English-speaking city by the early nineteenth century. Mid-twentieth-century rural-to-urban migration—accompanied by a modernization of economic life and a secularization of cultural life that became known as Quebec's "Quiet Revolution"—once again transformed Montreal into a majority French-speaking city by the 1960s.[29]

Montreal's Canadians of British Protestant heritage long retained control over senior management positions, especially in the private sector. An increasingly restive rising francophone middle class and university-trained intelligentsia demanded opportunities for promotion. Language served as the central battleground on which Montreal's and Quebec's French communities sought both a shared civic identity and greater access to the wealth of local society.[30] As Dominique Arel observes, "Young French speakers wanted to make careers in industry and finance, a milieu that was almost entirely English. At the same time, a spectacular decline in the birthrate of French-Canadians brought to the fore the question of the survival of the nation (culturally defined)."[31]

The stage was set for a formidable political movement embracing Quebec's sovereignty. With it began a Québécois drive to "'reconquer' the urban metropolis"—to make Montreal French-speaking once again.[32] Migrants from abroad found themselves caught in the middle of what would become a half-century-long contest that played itself out in Montreal's politics, economics, and culture, as well as in the city's spatial and physical development.

Historically, the principal north-south thoroughfare Saint Lawrence Boulevard / Boulevard Saint-Laurent demarcated a seismic fissure between these two parallel yet separate metropolitan communities (the failure to agree on a shared name for this important street suggests the tensions pulling Montreal toward separate French and English realities).[33] Montreal's neighborhoods came to sound different as language shaped the urban environment and to look different as various ethnic groups organized their lives and chose architectural forms that tied their Canadian present to some distant collective past elsewhere.[34] "Francophone" (native French-speaking), "anglophone" (native English-speaking), and "allophone" (migrants whose native language was neither French nor English) Montreal residents lived apart even as they were pushed together into sharing the same metropolis.

Bipolar French-English perceptions of reality remained intact despite the growth during the early to middle twentieth century of significant Irish

Catholic and Jewish communities (neither of which fit neatly into the English Protestant / French Catholic division of Montreal life). To provide some sense of the city's evolving ethnic mix, 97.5 percent of its population was of either French or British background in 1881, 93 percent in 1901, and 88.8 percent in 1911.[35]

By the 1980s, the percentage of the population of the Island of Montreal whose native language was either French or English (admittedly, a different measure) had fallen to about 80 percent.[36] These figures would continue to decline into the twenty-first century. The 2001 Canadian census revealed that approximately 80 percent of Montrealers still claimed either French or English as their native language.[37] However, only 44 percent of the city's population at that time identified themselves as being of French, British, or North American ethnic origin (a percentage half that which was reported ninety years before).[38]

This diversity was reconfirmed by a number of studies, including a subsequent survey of the parents of newborns in the city's maternity wards. According to research conducted in local hospitals throughout 2003, only 41 percent of the city's new parents planned to raise their babies as French speakers, as opposed to 25 percent who would use English with their children, and 34 percent who would speak to their offspring in some other language.[39]

However one may chose to calculate language use in the city, today's Montreal is a much more vibrantly multilingual town than the largely bilingual city of a century ago. Montreal's ethnic, religious, and linguistic composition had begun to change during the early twentieth century and accelerated after World War II with growing migration from Europe.[40] Portuguese, Greek, and Italian immigrants—like Jewish, Irish, and British immigrants earlier in the twentieth century—largely filtered into English-oriented Montreal.[41] Consequently, immigrants fell into the zero-sum language politics of the era as francophone Québécois increasingly categorized new arrivals from Europe together with English Canadians as "not us."

Some immigrants responded in kind, as is evident in the novelist and satirist Mordecai Richler's reminiscences about growing up in Montreal around the time of World War II. "Looking back," Richler would write during the mid-1960s, "I can see the real trouble was that there was no dialogue between us [Jewish immigrants and Québécois]." He continued:

Under the confessional system, we went to one set of schools and the French Canadians to another. I'm sure many of them believed that there was such an order as the Elders of Zion and that the St. Urbain Street

Jews were secretly rich. On my side, I was convinced all French Canadians were abysmally stupid. We fought them stereotype for stereotype. If the French Canadians were convinced the Jews were running the black market, then my typical pea-soup wore his greasy black hair parted down the middle and also affected an eyebrow mustache. His zoot trousers were belted just under the breastbone and ended in a peg hugging his ankles. He was a dolt who held you up endlessly at the liquor commission while he tried unsuccessfully to add three figures, or, if he was employed at the customs office, he never knew which form to give you. Furthermore, he only held this or any other government job because he was the second cousin of some backwoods notary who had delivered the village vote to the Union Nationale for a generation.[42]

MacLennan's character Athanase captured the unease on the other side of the Montreal linguistic divide. "The English were taking them [resources] over one by one," mused Athanase. "If the process continued indefinitely the time would arrive when the French in Canada would become a race of employees. Perhaps because they were a minority, perhaps because their education was not technical, they had no real share in their country's industry."[43]

In his memoir, Richler continued on to observe succinctly that "Montreal was, and remains, a sequence of ghettos."[44] The city was unquestionably Canada's most residentially divided city by the 1960s. Montreal's English and Jewish communities in particular inhabited the most segregated urban neighborhoods anywhere in the country well into the 1980s.[45] The 1981 Canadian census revealed, for example, that not a single Jewish resident lived in more than half the census tracts containing 53 percent of the city's residents, with nearly as high patterns of residential exclusion holding for the city's growing Indo-Pakistani and Chinese populations. About a third of Montreal's census tracts recorded no Greek or Portuguese inhabitants, while about an eighth had no black or Caribbean residents. Meanwhile, Jewish Montrealers lived in neighborhoods that were identified by census takers as being 70 percent Jewish, while Greek, Portuguese, and Chinese districts were nearly as segregated by ethnic origin.[46]

Three factors began to break down Montreal's linguistic frontiers during the last quarter of the twentieth century, producing the multilingual city of today: New migrants started to arrive from outside the North Atlantic region; Montreal—like all North American cities—spun off a massive suburbanized region as center-city residents from various socioeconomic, linguistic, and ethnic groups began to drive their automobiles out to new edge cities; and Bill 101, requiring immigrants to send their children to French-language schools, refocused the process of ethnic assimilation on francophone rather than anglophone urban institutions.

The impact of these factors will be explored in greater detail in the chapters to follow, which examine Montreal's capacity to adjust to the new realities of transnational diversity and explore how the city has created new diversity capital. As Annick Germain and Damaris Rose have observed, "The major transformations wrought by immigration waves in the sociocultural space of Montreal since the 1960s are leading to the substitution of the two solitudes image by a new one, that of the multicultural city."[47] This transformation holds clues to some of the ways in which diversity capital might be created.

## Washington: From "Chocolate City, Vanilla Suburbs" to "Rocky Road"

Persons of European heritage, primarily from the British Isles, came to farm along the Potomac River together with their African slaves not very long after the original French *habitants* began to settle the Saint Lawrence River Valley far to the north.[48] White plantation owners and black slave laborers arrived during the 1630s, setting out to build a reasonably prosperous plantation economy along the broad flatlands stretching around the river south of the Great Falls across to the prolific Chesapeake Bay.[49] The bustling tobacco port towns of Alexandria in Virginia and Georgetown—as well as the now-long-forgotten Bladensburg in Maryland—testified to their success. The scion of one such family would become the continent's most successful eighteenth-century general as well as the first president of the United States.[50]

This very same Potomac squire, George Washington, presided over the tumultuous debates leading to the designation of an area one day's ride from his own home plantation of Mount Vernon as the future nation's capital following particularly bitter disputes between Northern and Southern congressional delegations. The Residence Act of 1790 and its amendment in 1791 traded Northern acquiescence to abandoning the country's early Northern capitals in Philadelphia and New York in exchange for Southern payments on the growing national debt.[51] The state legislatures of Maryland and Virginia ceded 100 square miles of territory around the confluence of the Potomac and Anacostia Rivers to congressional jurisdiction in accordance with article 1, section 8, of the 1789 U.S. Constitution, which called for the creation of a federal district to serve as the home to the newly formed federal government (the lands within the District of Columbia lying on the southwestern banks of the Potomac were ceded back to the Commonwealth of Virginia in 1846).[52] Washington appointed three commissioners—Thomas Johnson and Daniel Carroll of Maryland and David Stuart of Virginia—to oversee the formation of the new city.

Washington and his commissioners spent the spring of 1791 huddled in intense negotiations with local landowners, before arriving at an agreement to purchase the land necessary to set out a sufficiently grand government core around which the rest of the city would eventually grow. The final agreement, which was reached in a Georgetown tavern room now carefully preserved within the walls of Ukraine's embassy to the United States, called for a payment of $66.66 for each acre used for government construction. In one of the city's first (and perhaps most elegant) real estate scams, Washington and his commissioners convinced the local gentry to donate land that would be reserved for public use free of charge. What the plantation and other landholders evidently failed to grasp as they downed their tankards of ale was that the megalomaniacal plan for the Federal City drawn up by the French engineer Major Pierre L'Enfant on the basis of Louis XIV's design for Versailles set aside 3,606 of the capital's original 6,611 acres for public streets and avenues and for public spaces.[53]

This is the point at which most histories of Washington begin, with L'Enfant and the African-American surveyor Benjamin Banneker drawing imaginary avenues through the forests and marshes down from the proposed site for Congress on Jenkins Hill across Tiber and Goose Creeks to the banks of the Potomac some two miles away.[54] L'Enfant, a brilliant but imperious Frenchman, would soon be dismissed when he tried to ram a portion of New Jersey Avenue through the front door of Daniel Carroll of Duddington's planned plantation house (this particular Daniel Carroll being the uncle of Commissioner Daniel Carroll). As would be so often the case in Washington's history, the white L'Enfant's name was to be enshrined in the national memory but the black Banneker received virtually no credit for the city's strikingly beautiful and grand design until quite recently.[55]

Washington has often been presented through highly abstract, colorful, and picture-perfect maps ever since, representations that bleach out a much more complex reality on the ground.[56] L'Enfant's untimely dismissal reveals that the city's site was hardly a clean slate open to planners' and politicians' dreams. It had been inhabited for a century and a half by European settlers and their African chattel (to say nothing of the Powhatans and other Native American peoples for whom the region had been home for centuries before either whites or blacks had arrived). Between the vibrant trading towns of Georgetown and Alexandria stood twenty-two slaveholding farms and plantations. The sites that would be occupied by the U.S. Capitol, the White House, and much of L'Enfant's grand national Mall were under cultivation by black slaves when Washington and his commissioners negotiated land purchase prices in Georgetown. For example, the city's largest slaveholder, Notley Young (who owned 265 slaves in 1790), occupied the site near where the Jefferson Memorial would be built.[57]

Despite numerous riots and protests, the slave trade continued within the District of Columbia until the Compromise of 1850.[58] A total of 1,774 African-American Washingtonians were still held as slaves just over a decade later, in 1861, when the Rebel troops in Charleston fired on Fort Sumter, igniting the Civil War.[59] Once freed, Washington's African-Americans would be prohibited from owning certain businesses, working in a number of professions, and living in many of the city's neighborhoods for almost a century more.[60]

The Washington short story artist (and future Pulitzer Prize winner) Edward P. Jones captured the city's seemingly invisible but nearly impervious racial boundaries in his tale "The Store," set in the 1960s.[61] The main protagonist sets off from Sixth and O Streets, Northwest, for Georgetown, a trip of at most a couple of miles. "The next week," Jones's hero muses, "I took the G2 bus all the way down P Street, crossing Sixteenth Street into the land of white people. I didn't drive because my father had always told me that white people did not like to see Negroes driving cars, even a dying one like my Ford."[62] Like Montreal, Washington was a city of disconnected communities facing one another across the boundaries of Sixteenth Street, Northwest, and Rock Creek Park. These separate worlds met in distrust, if at all.

Such history—like the memory of conquest and subordination in Montreal —shapes a city's realities arguably more strongly than any line on a surveyor's map. Though seemingly far removed from the lives of the twenty-first-century transnational migrants who now inhabit the sprawling and booming metropolitan region, the city's history undergirds the baroque, sometimes pernicious, often brutal, and always complex set of mutual dependencies so innocently embraced under the phrase "Washington race relations." Race was entrenched as a defining characteristic of the area well before the first American president convinced a handful of representatives and senators in sweltering Philadelphia barrooms to move the new country's capital city to his own backyard.

The District of Columbia's history as being home to African-Americans, the city's proximity to Southern states that were already home to the nation's largest populations of African descent, the steady jobs offered by the federal presence in the city, and the accompanying persons of wealth who desired domestic service combined to make Washington a center of black life and cultural achievement.[63] More than 10,000 black freemen lived in the Federal District at the time of the American Civil War (1861–65), a number that had grown to about 190,000 by the beginning of the twentieth century.[64] If only 3.8 percent of the city's population had been black in 1800, this proportion would have grown to 20.6 percent by 1840 before falling to 15.0 percent at the outbreak of war.[65] Nearly 300,000 African-Americans lived in Washington by 1900, a number that would grow to more than a half-million (or about 70 percent of the city's population) by 1970.[66]

Poor African-Americans initially inhabited back alleys near their white employers.[67] They were joined by the burgeoning African-American middle classes and so-called gilded classes, which initially settled on the elegant streets of Le Droit Park around Howard University before decamping later in the twentieth century for the Gold Coast along the Sixteenth Street, Northwest, corridor.[68] Indeed, poor and wealthy African-Americans—as well as those of middling wealth—were rather evenly distributed within the city until World War I.

The starkly segregationist and viciously antiblack employment policies of Woodrow Wilson's presidency undercut the earning power of many within the African-American community. In combination with expanding suburbanization, Wilson's policies enforced the city's racial divide. Washington's relatively small but important German, Irish, Jewish, Italian, and Greek communities became firmly incorporated into the larger white city.[69] The city's African-American population became ever more isolated in a circle of cramped neighborhoods downtown running from the old Southwest neighborhoods south of the Mall, around Capitol Hill, and into the near-in neighborhoods along U Street, Northwest.[70]

A tiny "Chinatown" was set apart from both whites and blacks, initially at the foot of Capitol Hill along Pennsylvania Avenue, Northwest, near the site of today's Canadian Embassy.[71] Members of Congress and local whites were disturbed by the Chinese presence at such a prominent location. In yet another manifestation of the local dictum that "white is right," the local authorities evicted all Chinese businesses from Pennsylvania Avenue in 1929, forcibly relocating them to the vicinity of Seventh and H Streets, Northwest, in one of many recurring efforts to "refurbish" the "Nation's Avenue." Rather than constituting a beachhead of interculturalism in a racially divided city, Washington's nascent Asian community became tossed together in official policies with the much larger African-American society as "nonwhite."

Restrictive covenants on housing deeds and the unwillingness of local banks to loan money to African-Americans and Asians consolidated such racial boundaries. Although outlawed by the Supreme Court in 1948, restrictions prohibiting nonwhites from purchasing property and living in many Washington neighborhoods exerted informal effects well into the 1950s. The city's strictly segregated school system forced parents to chose their neighborhoods carefully. And whatever legal barriers and school restrictions African-Americans might overcome, banks' "red lining" (whereby loans in certain neighborhoods either were not made at all or were not made to blacks) ensured that Washington maintained its own local form of apartheid.[72]

The city's ingrained racial divide became a shared heritage for all Washingtonians. African-Americans were as dismissive of their white neighbors as Caucasians were of blacks. In his extraordinary novel of Washington manners set in the high African-American society of the 1920s, Edward Christopher Williams had his leading protagonist, Davey Carr, consider the mean-spiritedness of his fellow Washingtonians. "I guess an all-wise Providence knew what he was doing," Davey wrote to his best friend in New York, "when he evolved the Nordic type, with its watertight, noncommunicating compartments in morals and religion, but to me it still remains the greatest of all the riddles of humanity."[73]

Early-twentieth-century Washington could not last. Beginning in 1950, the city stood at the beginning of a stunning transition from a modestly sized, culturally Southern city to a major international metropolitan region. The middle decades of the twentieth century were marked by a dramatic expansion of Washington's size and significance within the United States. Beginning with the explosion of the federal government during the New Deal, and continuing through World War II and the Cold War, increasing numbers of Americans were drawn to the Nation's Capital, either to work for the federal government directly or to influence those who did.[74] The city's population increased from 663,091 in 1940 to a high of 802,178 in 1950, before declining to 756,510 in 1970 and 572,051 in 2000.[75] The population of the

Washington metropolitan area grew from 967,985 in 1940 to 2,860,510 in 1970 and to more than 7 million by 2000 (by which time the U.S. Bureau of the Census maintained that, for statistical purposes, the Washington metropolitan area had merged with that of nearby Baltimore).[76]

During the 1970s, Washington's racial realities were sometimes portrayed as an ice cream cone on which sat a chocolate scoop of ice cream pushed down by more plentiful portions of vanilla ice cream. This image no longer has validity. Clusters and chunks of an increasingly diverse population are to be found throughout an expanding metropolitan region. Given the centrality of the automobile to the current Washington lifestyle—and some of the longest commuting times in the country—the region increasingly resembles a cone of the composite ice cream flavor "Rocky Road." Washingtonians, in the process, have created new diversity capital for themselves and for their region in a metropolitan area that had been more noted for racial barriers and hostility.

## Kyiv: Dividing Lines That Are Not on Any Map

Kyiv lies much closer to the twentieth century's heart of darkness than either Montreal or Washington. The city, which was fought over by Whites and Reds during the Russian Revolution, was incorporated into fourteen to twenty different "regimes" between 1917 and 1920. Much of Kyiv was destroyed during World War II, when the city witnessed some of the worst atrocities of the Holocaust and of Stalinist repression. As a consequence of such turmoil, the ancient Ukrainian capital is, in fact, a demographically much newer city than either Montreal or Washington. Visitors to all three cities would have a much easier time identifying descendants of early-twentieth-century Montrealers or Washingtonians in their cities than finding a Kyivan with an ancestor who resided there before World War I.

Kyiv's linguistic and religious divisions date back centuries, to times well before either Montreal or Washington could be found on maps. One of the many significant consequences of the dramatic disjuncture between historical and contemporary Kyiv is that the sort of group demarcations so physically present in Montreal and Washington do not appear to exist in the Ukrainian capital. The rupture between Ukrainian and Russian linguistic communities—and among Moscow Patriarchate Orthodoxy, Kyiv Patriachate Orthodoxy, Autocephalous Orthodoxy, the Greek Catholic (Uniate) Church, Roman Catholicism, Islam, and Judaism[77]—cannot be identified on city maps with the certainty of a Saint Lawrence Boulevard / Boulevard Saint-Laurent or a Rock Creek Park–Sixteenth Street, Northwest, corridor. Yet though the linguistic and cultural tensions between Ukrainian and Russian may appear to be less acute than either linguistic or racial conflicts in Montreal and Washington, very real contradictions are to be found within the hearts and minds of tens of thousands of city residents.

The last Soviet census in 1989 provides a small glimpse at some of the complexities of identity in contemporary Kyiv. Soviet census takers reported that, in 1989, 72.4 percent of the city's 2,595,000 residents were ethnic Ukrainians, 20.9 percent ethnic Russians, and 3.9 percent ethnic Jews, with a scattering of Belorussians, Poles, and others.[78] Though a vast preponderance of Kyivans around the same time reported themselves to be bilingual in Ukrainian and Russian, many ethnic Ukrainians claimed Russian as a native language.[79] Thus, ethnic and linguistic identity did not coincide. The city, like Ukraine itself, appeared to be a very different place and society depending on whether one chose ethnicity or native language— let alone religion—as the principal indicator of identity.[80]

The 2001 Ukrainian census revealed a significant increase in nationality identification as Ukrainian throughout the city (82.2 percent of the city population telling census takers that they were Ukrainian, 13.1 percent reporting to be Russian, and only 0.7 percent claiming to be Jewish).[81] Contemporary Ukraine's—and Kyiv's—uncertainty about identity continues, rooted as it is in the extraordinarily complex histories of the people who have inhabited the region's lands.

Kyiv's multiple ambiguities may be traced throughout the city's centuries-long history. Its founders built one of Europe's most powerful cities, taking advantage of its position at the crossroads of east–west trade routes between Central Asia and Europe and a remarkable north–south river system, which permitted navigation between the Baltic and Black Seas. Ancient Kyiv, and the principality ruled from the city, were significant places by the mid–ninth century, when Varangians entrenched in Kyiv continued on to attack Constantinople with 200 ships.[82]

Grand Prince Volodymyr "the Great" Christianized the Rus' in 988. He was followed by perhaps Kyiv's most effective ruler, Grand Prince Yaroslav "the Wise," who between 1019 and 1054 controlled vast territories extending north to the Baltic and northeast into what is today central Russia. In 1037, Yaroslav oversaw the creation by Constantinople of a metropolitanate for all the Rus' based in Kyiv, thereby making the city a major ecclesiastical center.[83] The prestige of Yaroslav's court was reflected in the marriage of his daughter Anna to Henry I of France.[84]

Destructive internecine dynastic warfare following Grand Prince Yaroslav's reign dramatically weakened the Kyivan state, which would eventually collapse as the Mongols sacked the city for the first time in 1240.[85] Kyiv would be destroyed again in 1416 and 1482, further separating the city's present from its past.[86] Large parts of today's Ukraine fell to the Polish-Lithuanian Grand Duchy between the thirteenth and mid-seventeenth centuries, before the Cossack Hetman Bohdan Khmel'nyts'kyi signed the Pereiaslav Treaty with Muscovy's Aleksei Romanov to form an anti-Polish alliance in 1654.[87] Eastern Ukraine and the regions around Kyiv came ever more strongly under Moscow's (and Saint Petersburg's) dominance during the next three centuries.[88]

Heated debate still rages over this early history and over the precise identity of the Rus', as well as over the significance of the state that emerged.[89] Were the Rus' Normans? Scandinavians? Were the Varangians outsiders who held colonial sway over the local Slavic tribes that had traveled to the region from Central Asia? Did the years of chaos following the collapse of the Rus' state produce three distinct East Slavic peoples—the Ukrainians, the Russians, and the Belarusians? Alternatively, are today's Ukrainians the true inheritors of Rus' glory? Is there a direct link through Yaroslav the Wise from Kyivan accomplishment to Muscovy, thereby making today's Russians claimants to Kyiv's history? Is Kyiv "the Mother of all Russian Cities," or are the Russians merely a lost Slavic tribe from Bulgaria?

Such questions bedevil Russian, Ukrainian, and Belarusian twenty-first-century state builders; how one chooses to respond to such questions helps to define one's identity.[90] Herein lie the origins of the internalized ethnic and linguistic fault lines that run through today's Kyiv. The interpretation of such events will help to determine how much diversity capital Kyiv will generate in the years ahead.

Beyond questions of ideological and historical debate at present, Kyiv's urban reality after the 1654 Pereiaslav Treaty proved to be distinctly less than grand until well into the nineteenth century. The city's decline occurred even though Kyiv enjoyed the communal and commercial rights granted by the Magdeburg Charter from the fifteenth century. If, for example, 30,000 souls inhabited "Old Kyiv," Podil', and Pechersk during the twelfth century, that number fell to under 9,000 by the fifteenth century. Subsequent growth well into the eighteenth century took place in the commercial waterfront lowlands of Podil' and the monastic highlands of Pechersk rather than in more centrally located Old Kyiv.[91]

Kyiv's fate would improve during the rule of Russian empress Catherine the Great as she introduced a new Town Charter in 1785, after having turned the city into a main supply base for the Russian army during the Russo-Turkish War of 1768–74.[92] Lorded over now by police and senior government officials appointed from Saint Petersburg, Kyiv became the administrative center for territories taken by Russia from Poland during the second partition of that country.[93] Saint Vladimir's University opened its doors in 1834, just as Nicholas I eliminated the last vestiges of the city's Magdenburg communal rights. Consequently, the main pillars of a new Kyiv—religious institutions (dating from the times of the Rus')), trade, defense industries, education, imperial administration—were in place by the mid–nineteenth century. The eighteenth century city of about 15,000 had grown to nearly a quarter of a million by the time of the 1897 Imperial Russian census.[94] Kyiv entered the twentieth century with a population 57 percent of whom reported Russian as their native language, 23 percent Ukrainian, 13 percent Yiddish, and 6 percent Polish, with small groupings of residents who identified other languages as native (*rodnoi iazyk*).[95]

The Kyiv of the 1897 census would be destroyed many times over during the subsequent half-century. The city was the site of heavy fighting and fierce pogroms in the years around 1905.[96] As the capital of a short-lived Ukrainian republic, Kyiv became a focal point of warfare between the Bolsheviks and their numerous opponents between 1917 and 1920.[97] Because the Soviets were uncertain about the city's loyalties, they established the capital of their new Ukrainian Soviet Socialist Republic in Khar'kiv, which quickly eclipsed Kyiv as Ukraine's largest industrial and urban center. The Famine of 1932–33, which devastated the Ukrainian countryside, inflicted heavy losses on Kyiv as well.[98] NKVD investigators took a particular vengeance on the city during the Stalinist purges of the 1930s, after the city had become capital of Ukraine once again in 1934.[99] Revolutions, bloody pogroms, world wars, civil wars, purges, and collectivizations were arguably as destructive to the city that had taken shape in the nineteenth century as the Mongol conquests had been centuries before. Then came the Nazi occupation of World War II.

German rule proved to be especially brutal in Kyiv, as testified to by the horrific slaughter of 34,000 Jews on the night of September 29–30, 1941, at the Babi Yar / Babyn Iar ravine.[100] More than 100,000 Kyivans in all were murdered at this site. In an act that can only be considered loathsome even by the grotesque standards of high Stalinism, local administrators temporarily converted Babi Yar into an amusement park immediately after the war.

Retreating Nazi and advancing Soviet armies had devastated the city by 1943, and most of its bridges and its main avenue—the Khreshchatyk—lay in ruins.[101] The city slowly rebuilt even as fierce partisan wars continued into the early 1950s in recently Sovietized Western Ukraine, as well as closer to Kyiv between Ukrainian and Polish guerrilla brigades.[102] The 1960s and 1970s were a time of dramatic migration into the city from the surrounding countryside, of purges and arrests of Ukrainian "nationalists" both within and outside the Communist Party, and of sustained Russification campaigns in every aspect of Kyivan life.[103] These events were capped off in 1986 by the Chernobyl' nuclear reactor accident just to the northwest of the city, which is now known to have released dangerous levels of radiation over Kyiv.[104]

The traumatic November 2004 Ukrainian presidential election, subsequent protests (which drew perhaps as many as 1.5 million demonstrators onto Kyiv's streets), and the accompanying "Orange Revolution" exposed the extent to which many of the divisions inflicted over the course of the twentieth century had yet to heal by the beginning of the twenty-first century. Each new episode left fresh and deep unhealed scars cut deep into the Kyivan psyche.[105]

Taken together, such events have made Kyiv a much less stable and fully formed urban community than either Montreal or Washington in that there is, at one and the same time, far too much history and far too few people who actually have experienced it. The need for new diversity capital is acute. Only time will reveal whether or not the residents of the newly independent Ukraine master the skills necessary to accommodate a great diversity of people.

Little is particularly new or noteworthy about the stories of individual migrants in each of these divided metropolitan regions, no matter how compelling a particular life may be. For example, if one considers contemporary migration to Kyiv, a vast majority of the city's transnational migrants during the late 1990s and early 2000s made their way via age-old patterns based on contacts with friends and family members who had become residents.[106]

Yet the present-day story of transnational migration to Kyiv remains important for several other reasons. First, Ukraine has been suffering demographic decline, as have many of its formerly Soviet neighbors. Overall, Ukraine's population has fallen from nearly 52 million when it became independent in 1991 to less then 48 million in 2004. This loss is due to many factors, including high levels of age-specific male mortality, a generally aging population, a falling birthrate, and out-migration to Russia as well as to Western countries such as Portugal, Canada, England, and the United States.[107] A small but telling indicator of the depth of Ukraine's demographic dilemma is that the number of schoolchildren entering the first grade on September 1, 2004, was only 40 percent of that on September 1, 1991.[108] Because of these dramatic reductions, Ukraine must become a nation of migrants to sustain anywhere near its population at the time of independence.

Second, migration has already begun to have an impact on Ukraine in general, and especially on Kyiv. In August 2004, Serhiy Brytchenko, the head of the presidential administration's Migration Directorate, called attention to increasing migration into Ukraine in an article appearing in *Uryadovyy Kuryer*, the newspaper of the Ukrainian Cabinet of Ministers.[109] Brytchenko reported that the number of people granted Ukrainian citizenship increased by 40 percent during the first six months of 2004 in comparison with the similar period in 2003, and by 180 percent in comparison with the first half of 2002. The Ministry of the Interior similarly observed an increase in applications for permanent residence, the initial step toward Ukrainian citizenship.

Brytchenko continued on to note that the number of people adopting Ukrainian citizenship substantially surpassed the number relinquishing citizenship. Ukraine, he concluded, is becoming increasingly attractive to foreigners and persons without citizenship. In short, it is becoming a country of immigration for the first time in recent history.

Calculations of the number of transnational migrants living in Kyiv in particular are notoriously unreliable, with strikingly different data becoming available from any number of state agencies. Overall, 101,268 foreigners were registered with municipal and district passport registration and immigration authorities throughout Kyiv in 2001, when the city's official population stood at 2,606,716.[110] Such numbers—though modest in comparison with those of the Montreal and Washington metropolitan areas—represent a significant increase from the Soviet period. An examination of transnational migrants in Kyiv thus enhances an understanding of developments within Ukraine.

Third, and more important for readers concerned with migration issues generally, transnational migrant communities appear to be taking root in Kyiv.[111] In other words, this is a moment of creation for a migrant metropolis. These trends can be seen in a city and a country that are struggling to establish a new democracy and a new state. Consequently, an examination of the impact of transnational migrants on postindependence Kyiv highlights the potential relationships among migrants, nascent democratic institutions, and diversity capital. Kyiv thus provides an invaluable counterpoint to discussions of more established migrant cities and democratic regimes in North America such as Montreal and Washington.

Migrant communities are making their presence felt in Kyiv, forcing local institutions to adapt and long-standing authority relations to change. Overall, the level of migrant interaction with native residents—Ukrainian, Russian, Jewish, and others—varies considerably from one migrant community to the next. Vietnamese, Iraqi, and Pakistani migrants, for example, report having both work-related and leisure time contact with the local Kyivan population on a routine basis. Afghan and some African migrants, by way of contrast, deal primarily with one another.[112] Moreover, many of Kyiv's transnational migrants live in relatively self-contained communities in patterns not dissimilar from those of migrants in many other cities and metropolitan regions around the world.

As is often the case for ethnic communities seemingly everywhere, nearly all transnational migrants to Kyiv maintain traditional dietary and dining patterns and customs. Only a few attempt to sustain their native languages in any organized manner (though most speak them on a daily basis). Kyiv's migrants may be unusual, however, in that they tend to be secular no matter what their previous religious background. In surveys, migrants to the Ukrainian capital rarely claim to have made a special effort to sustain the religious culture and practices of their homeland.[113]

## Learning from New Stories

Overall, with the exception of what transpires in kitchens and dining rooms, Kyiv's migrants appear to meld with and enrich the practices and customs of the city's mainstream even as they live primarily among themselves. Yet pressing questions remain: Which language community's mainstream— Ukrainian or Russian? And to what extent are the migrants reshaping the host culture in return?

Washington offers a possible point of comparison. If Kyiv conforms to familiar spatial and social patterns for migrant assimilation or nonassimilation within a new political context, the Washington metropolitan region suggests new spatial relationships between migrant and host society that are made possible by cyberspace and the automobile.

The U.S. capital and its surrounding metropolitan region emerged during the 1990s as the fifth largest recipient of legal transnational migrants among the metropolitan regions of the United States.[114] Washington's large professional job base began to attract ever more educated new residents, including affluent minorities and transnational migrants. Like many American metropolitan areas, the Washington-Baltimore region had been losing residents to the Sunbelt. The region's population nonetheless continued to experience overall population growth as a result of the arrival of hundreds of thousands of migrants from other countries. Metropolitan growth has rested to a considerable degree on young college graduates and other well-educated African-Americans, Asians, and Hispanics who are drawn to the area's high-end labor market.[115]

In other words, the Washington region has become an "immigrant magnet metropolis."[116] The region is increasingly becoming home to substantial proportions of various segments of the U.S. transnational migrant population. For example, the 100,000 or so African-born residents of the Washington region represent approximately one-tenth of the total U.S. population of Africans.[117]

As in Kyiv, interaction among migrants and more traditional Washingtonians varies according to ethnicity and income level. Central American migrant communities appear to remain more self-contained, both spatially and socially, than other new arrivals. Many migrants from a variety of backgrounds have fanned out across an expanding metropolitan region that extends from Frederick, Maryland, in the west to the shores of the Chesapeake Bay and beyond in the east; and from north of Baltimore to south to Fredericksburg, Virginia, a hundred-plus miles away.[118] The Washington region's dispersed residential patterns appear to have integrated newcomers into the area's life more seamlessly than have the patterns of Kyiv, where transnational migrants are more and more confined to a few neighborhoods.

The diversity of Washington's migrant settlement patterns ameliorate the isolation of various migrant communities. Washington and its region have become home to an expanding migrant population in which no single group dominates.[119] Not very many Americans—or even Washingtonians—appear to be aware of the new metropolis that has taken shape so quickly during the past quarter-century. The image more often used to discuss the capital metropolis remains one of "black" and "white," even as metropolitan reality has become something quite different.

Montreal, by contrast to both Kyiv and Washington, has been a city of transnational migration almost from its founding in 1645.[120] New arrivals since the early 1960s have made Montreal into quite another place, landing in planes on flights originating largely in Asia, Africa, South America, and the Caribbean.[121] Montreal is no longer a city of French and English alone, but one with distinct communities of color. According to some estimates, metropolitan Montreal's population of so-called allophones (a group that also includes individuals who are not transnational migrants) has now surpassed that of anglophones.[122]

The absorption of such "new immigrant" groups into Montreal's linguistically divided cityscape has occasionally been difficult.[123] The city's numerous ethnic communities have long remained both spatially and socially distant from one another.[124] Its traditional linguistic divide appears, at times, to continue to influence the behavior of migrant groups that might otherwise have more in common with one another than either has with native inhabitants. Haitian and anglophone West Indian immigrants, for example, rarely interact with one another despite their shared racial and geographic origins. Rather, Haitians settle in the francophone community of the city's east end while West Indians move to the anglophone west. Linguistic affinity with the city's two charter groups overrides whatever other racial, social, economic, or cultural affinities these former Caribbean islanders may share.[125] As such trends suggest, the realities of everyday life in Montreal, as in Washington and Kyiv, have become ever more complex during recent decades.[126]

## Creating Diversity Capital?

The growing presence of individuals who do not fit into long-standing group boundaries fundamentally alters the social, cultural, and political contours of traditionally bifurcated metropolitan regions such as Montreal, Washington, and Kyiv. What happens when migrants gain a significant foothold in an environment thought of in starkly binary terms? How does that presence change perceptions and institutions? How is diversity capital created and conserved? Such questions stand at the center of this study.

This volume attempts to confront such issues by exploring the impact of migrants on three urban communities that represent different stages in a cycle of urban change whereby large numbers of transnational residents transform previously "divided" urban cultural, social, economic, and political landscapes. This pattern—which may also be identified in such varied cities as Barcelona, Brussels, and Osaka—is perhaps most mature in Montreal and may be said to have already profoundly altered how that city functions. Washington represents a city where demographic transformation has been largely unnoticed and is only now emerging into general public awareness. The processes examined in this volume are only in their infancy in Kyiv.

Chapter 2 begins this inquiry by identifying residential patterns for migrants in each city. Chapter 3 examines local labor markets and educational institutions in all three cities so as to better explore how migrants in these divided cities make ends meet and adapt to their new surroundings. Schools—from kindergartens through universities—serve as major arenas where migrants and representatives of more traditional competing groups come together to create new diversity capital, as do workplaces.

The study concludes with three chapters that retell the stories of the complex political transformations that all three cities have experienced at the same time as transnational migrant groups have been taking up residence in their new homes. Montreal, Washington, and Kyiv sustain very different political regimes now than they did four or five decades ago. Transnational migrants are important contextual factors for these changes, even if they often seem only tangential to many of the dramatic political conflicts of recent years. In other words, migrants are present whether or not they appear in every paragraph or on every page of these chapters.

Such accounts of local politics demonstrate how the diversity capital generated by the accommodation of transnational migrants in neighborhoods, schools, and workplaces can become a source of both sustenance and mischief in urban political struggles. Political leaders in Montreal have drawn on their city's expanding base of diversity capital to redefine a metropolitan-scale community of cosmopolitan cultural complexity over a period extending from Saint Jean-Baptiste Day in 1968 to Saint Patrick's Day in 2004. In Washington, enhanced diversity capital has constituted a resource in reinforcing a post–civil rights municipal regime of neighborhood-oriented problem solving. And Kyiv's post-independence city builders can either draw on transnational communities to define an inclusive post-Soviet identity for their metropolitan region or can seek out new pariahs against whom to delineate precisely what it means to be "Ukrainian." One way or another (most often, perhaps, implicitly rather than explicitly), the growing presence of thousands of residents who have not experienced Montreal's or Kyiv's historic linguistic divisions—or Washington's tortured racial relations—is reshaping the rules of political life in these three metropolitan regions.

The capacity of divided cities such as Montreal, Washington, and Kyiv to absorb migrants—their diversity capital, as it were—may initially appear to be but a recent version of the long-known saga of newcomer and unfamiliar town confronting one another. However, this volume suggests that there is a qualitative difference in many of today's migration stories, because transnational migrants constitute cultural and religious communities that have not traditionally been associated with their new places of residence.

To return to where this introduction began, the more organic accumulation of diversity capital in entrepôts such as Amsterdam and New York offers instructive, yet incomplete, tales of how communities can accommodate difference. By contrast, the world's divided cities may well reveal a more fulsome range of options open for urban life in the decades ahead.

# Part II

# Social and Economic Transformations

# Chapter 2

# Living in the Middle

It is not only what we have inherited from our fathers and mothers that exists again in us, but all sorts of old dead ideas and all kinds of old dead beliefs and things of that kind. They are not actually alive in us; but there they are dormant, all the same, and we can never be rid of them.

—Henrik Ibsen, *Ghosts,* Act II, 1881

Historically, individual urbanites have sought out neighbors whom they see as similar to themselves. People seem to feel more at home living among others like themselves. Such group clustering has created identifiable districts in which various racial, religious, ethnic, and linguistic groups become segregated from one another—at times by choice, but all too often by force of law, custom, and banking practices. The result are cities transversed by invisible boundary lines that divide urban space into compounds impenetrable by the "other."

Migrants from abroad may have difficulty accommodating themselves to residential patterns created with others in mind. They often come to live in the middle, between groups, rather than becoming integrated into a more encompassing general reality. This has been the case in Montreal, Washington, and Kyiv, three cities in which new, more diverse groups of migrants began to arrive late in the twentieth century. Not quite conforming to traditional local linguistic and racial categories, tens of thousands of recently arrived Montrealers, Washingtonians, and Kyivans have transformed their neighborhoods into newly intercultural communities that are creating diversity capital.

The latest waves of transnational migration correspond to the rapid expansion of all three cities into massive metropolitan regions. Suburbs on the metropolitan fringe have, at times, proven themselves to be more open to newcomers than long-entrenched traditional inner-city neighborhoods.[1] Migrants from abroad and real estate developers (including state construction trusts) have reinvented all three cities during the past quarter-century, making them each far different places than they were in the past; and far more complex than longtime residents and politicians appear willing to accept.

## Montreal: A French Cosmopolitan Metropolis

Montreal has evolved during the past third of a century from a city divided between two founding linguistic and cultural communities—French and English—into a complex intercultural metropolis rooted in a generally shared knowledge of French. This dramatic deviation from past patterns in every aspect of the city's life has become especially visible in the spatial distribution of housing. To appreciate how this is so, it is important to begin with a review of the evolution of migration patterns into Canada, the Province of Quebec, and the Montreal metropolitan area during the past three decades.

Recent migration within Canada has changed along three dimensions, which have dramatically reshaped the urban experience of Montreal as well as public discussions of Canada's and Montreal's new realities.[2] First, transnational migrant flows have shifted from Montreal to Toronto and the Canadian west as the economic uncertainties generated by debates over Quebec's sovereignty have devalued the Montreal region as a receiving area for migrants from abroad. Second, migrants to Montreal—as everywhere in Canada—have increasingly arrived from the Caribbean Basin, Africa, Asia, and Latin America. Third, Montreal's migrants have more and more come from the francophone worlds of Africa, Haiti, and Southeast Asia.

As late as 1985, some researchers still referred to Montreal as "Canada's second largest center of immigration."[3] Though the Montreal metropolitan region continued to collect the vast majority of migrants to Quebec, the province's overall share of migrants officially arriving in Canada began to fall precipitously. Montreal's failure to attract transnational migrants relative to Toronto and Vancouver had become undeniable during the 1980s. By 2001, 43.7 percent of the population of metropolitan Toronto was foreign born, as was 37.5 percent of metropolitan Vancouver's population. Both figures stand in stark contrast to the 18.4 percent of the residents of metropolitan Montreal (and only 2.9 percent of the residents of the Quebec City region) who were born outside Canada.[4]

However, the actual size of Montreal's transnational migrant community appears to be significantly larger if one includes the children of migrants born in Canada. Such a recalibration of census data for Montreal suggests that the city's "extended immigrant population" has remained larger and younger than official data might suggest.[5] Nevertheless, the basic national pattern of Montreal lagging behind Toronto and Vancouver as an transnational migration destination has persisted for a generation.[6]

Signs of change began to appear during the 1990s, when Quebec's share of "refugee claims made on Canadian soil" rose even though the province's overall share of total Canadian migration stood at only 16 percent throughout the decade.[7] A threshold of sorts was passed in 2002, when for the first time in a decade Montreal drew more transnational migrants (33,004) than Vancouver (29,922).[8] Though outpacing all other Canadian cities but one as a transnational migrant destination, Montreal nonetheless lagged far behind Toronto, which captured nearly half of all transnational migrants to Canada that year (111,580, or 48.7 percent of the national total).[9]

Increasing transnational migration to Montreal during the first years of the twenty-first century became essential to the city's well-being, given otherwise precipitous projected declines in its total population and labor force.[10] The upturn evident by 2002 marked something of a success for aggressive provincial policies and practices intended to attract more of Canada's newcomers to Quebec.

Prosovereigntists in Quebec City had become concerned by the prospect of their province's ever shrinking proportion of the total Canadian population by the 1970s. If, at the time of Canadian Confederation in 1867, Quebec's population accounted for something approaching half the country's residents, this number had fallen to less than a quarter by 2000.[11] The share of French speakers within Canada similarly began to decline in the middle decades of the twentieth century.

Successive Quebec governments supported demographic and linguistic research to delineate these new realities through grants as well as the work of quasi-government organizations such as the Council of the French Language. These efforts lent authority to the notion that the proportion of the overall Quebec population for whom French was the maternal language had entered a period of decline around 1950.

Although this trend slightly reversed itself during the early 1980s, many concerned over Quebec identity envisioned a tidal wave of anglophone transnational migrants overwhelming their tiny francophone corner of North America.[12] Historically, Native Québécois had long been concerned about workers arriving from abroad. Hostility had grown during the previous century as rural French Canadians were emigrating to New England even as Europeans were arriving to work in the province's expanding urban industrial centers.[13]

Within this broad historic context, the future of the francophone majority on the Island of Montreal appeared to many Québécois observers to be especially bleak. Quebec nationalists expressed increasing alarm over the prospect of a francophone minority within the metropolitan Montreal region.[14] As the sovereigntist (*souverainiste*) leader René Lévesque would observe in 1968:

> We need only glance at the school situation in Montreal: in 1965, children of immigrants to Quebec made up thirty-five per cent of those attending Anglo-Protestant schools, sixty-four per cent (mainly Italians) of those attending Anglo-Catholic schools, and barely 3.5 percent of the 170,000 children in French Schools! . . .
>
> There is no place in the world where immigrants move spontaneously to join the minority, unless that minority is a dominant one.
>
> But this is enough to see clearly where we stand and also what we must do, and do quickly. In a few years, it will be too late.[15]

In a historic turnaround, Quebec's nationalists began to see the potential linguistic benefit of welcoming migrants from abroad to the province. The decline in the province's birthrates was too rapid to ignore. Quebec, which had the highest birthrate in Canada as late as 1959, entered the 1970s with the nation's lowest rate of natural population replacement.[16] This drop was especially pronounced among francophones, raising the specter of an emerging anglophone Quebec by default. French linguistic dominance of the province in general, and of Montreal in particular, could only be sustained by the arrival of francophone migrants from abroad. Given such demographic realities, various Quebec governments began to look with favor on transnational migration as a way to replenish the province's population.

This shift from suspicion of transnational migrants to embrace took place just as Canada began to attract a new kind of migrant. As recently as 1957, 95 percent of migrants to Canada from abroad traveled from Europe or the United States, and migrants from the United Kingdom constituted the single largest group of new arrivals. In 1990, however, migrants from Europe (including the United Kingdom) and the United States represented only 29 percent of new arrivals to Canada—as opposed to 49 percent from Asia.[17]

This general change in migration patterns was evident in Quebec as well.

During the 1970s, Haitians represented the largest community of newly arriving migrants to the province (and, essentially, to the Montreal metropolitan region). They were followed by migrants from the United States, France, and Vietnam. Haitians remained the largest group of new migrants to Quebec during the 1980s, with settlers from Lebanon, Vietnam, and France forming the next largest groups.[18] One consequence of these trends emerged in the 2001 Canadian census count, which revealed that "visible minorities" had come to make up nearly a fifth (18.7 percent) of the city's overall population.[19]

Within less than two generations, Montreal's transnational migration community had been transformed from being largely anglophone to increasingly francophone, and from being predominately of European heritage to constituting an ever more visible minority. These communities did not fit comfortably into the century-long pattern of largely non-French-speaking migrants flowing through the English-speaking West End. Initially settling along the various interstices of the Saint Lawrence Boulevard/Boulevard Saint-Laurent corridor, new migrant communities spread into previously Québécois neighborhoods further east. Their arrival coincided with a period of dramatic suburbanization, which broke down the harshly bifurcated linguistic residential patterns of previous decades.[20] Non-European francophones were beginning to convert Montreal into a French-speaking interethnic metropolis.

The onset of late-twentieth-century North American–style suburbanization has played a transformative role in Montreal's spatial composition. An older North American city, Montreal evolved from a walking town to a streetcar city during the late nineteenth and early twentieth centuries. The city emerged from World War II as a relatively centralized urban settlement spread out along rail and trolley lines focused on the Island of Montreal. As elsewhere in North America, the postwar period would be one of rapid expansion along new highways. The geographer Jean-Claude Marsan perceptively captured this change in his 1981 exploration of the city's physical development. Writing with words and images that would apply to nearly every North American city of any consequence, he observed:

> Today, we see a fragmented metropolis pulling towards the undefined fringe of its own perimeter which is gradually but steadily eroding the rural countryside. It is hard to detect any kind of physical or social framework on the periphery, even the linguistic dividing line no longer follows the pattern of the previous century. The only remaining boundaries are

those that vaguely separate the various social classes. In fact, the built-up expanses are no longer a city in the real sense of the word, but an urbanized area seemingly devoid of any spirit or identity except for a few square miles at the centre of the city and in its immediate neighborhood which have retained some of their traditional aspects.[21]

In the years around the time that Marsan was writing his study of the region's spatial evolution, francophones and allophones were moving out of the central city to other areas of the Island of Montreal, and beyond. During the early 1980s, the populations of both groups were growing faster on the fringe than in central neighborhoods (the anglophone community was in decline everywhere throughout the region).[22] Anne-Marie Seguin and Annick Germain would write, at the end of the century, that Montreal had spread "out beyond the original Island of Montreal to create a polynucleated urban agglomeration. Greater Montreal is a city region of, in 1996, 3,326,510 inhabitants. It is highly fragmented with 113 municipalities of different sizes each defending its own autonomy."[23] The central city of Montreal accounted for only about one-third of the region's overall population.

In another development foreshadowed by Marsan, income rather than language was emerging as a more powerful determinant of where one lived within Montreal. The region's deindustrialization reached full fury during the 1970s and 1980s, devastating working-class neighborhoods in central Montreal. In a quarter-century, the manufacturing area along the Lachine Canal—once Canada's most highly concentrated industrial district—lost half its population and nearly all its jobs.[24] Because of this decline—in combination with the economic decline linked to political uncertainties arising from the mounting campaign for Quebec sovereignty—the Montreal region would enter the 1990s with one in five Montrealers residing with other family members, and one in two unattached individuals, living in poverty.[25]

The seemingly renewed prosperity of the 1990s did little to alleviate these broad patterns. The 2001 Canadian census revealed that Montreal retained the highest incidence among all major Canadian cities of single-parent families (21.1 percent) and one-person households (37.6 percent), just as had been the case a decade before.[26] Montreal similarly continued to suffer from the highest unemployment rate of all Canadian large cities (10 percent in 2003), the highest incidence of low-income households (22 percent in 2001), and the highest proportion of residents receiving some government assistance (14 percent in 2001).[27] More than a third (37.8 percent) of Montreal children under six years of age lived in poverty in 2004. This figure was less than the 44.4 percent of 1995 but more than the 35 percent poverty rate among children in this age group in 1990.[28] Poor Montrealers are increasingly concentrated in the central city in a pattern more superficially reminiscent of the United States rather than of large Canadian cities to the west.

Creative municipal housing programs—such as Operation 10,000 Logements, launched in 1979—began to repopulate abandoned manufacturing and warehouse districts with a rising (and largely francophone) middle class. As a consequence of these government interventions into the housing market, both middle-class and impoverished Montrealers were scattered in an ever-more-checkerboard-like pattern across the city and the metropolitan region.[29]

Montreal's residential pattern became increasingly complex. On the one hand, groups such as asylum seekers drew on a variety of government programs to find housing in mixed-income neighborhoods of their corresponding ethnic group.[30] On the other, transnational migrants benefited from extensive social networks in Montreal and elsewhere in Canada as they sought to resolve their housing and other challenges in their new country.[31]

Montreal's leading transnational neighborhoods—Côtes-des-Neiges and the neighboring Nôtre-Dame de Grâce—reflect this new metropolitan-scale socioeconomic, ethnolinguistic spatial amalgam.[32] At times known as Montreal's "Bronx," Côtes-des-Neiges has become home to Africans, Arabs, Cambodians, Jews, Filipinos, Lao, Vietnamese, Chinese, Latin Americans, Portuguese, Haitians, and members of various other groups living in close proximity to one another.[33]

The area existed as a small rural village from the seventeenth century through to the twentieth, when it began to grow slowly.[34] By World War I, Côtes-des-Neiges was brought within Montreal's official boundaries under one of the city's two dozen annexations of surrounding communities. Major cultural institutions followed, including the Oratory of Saint Joseph, Brebeuf College, the Jewish General Hospital, and, in 1943, an expanded University of Montreal. Correspondingly, the area's population grew from 2,444 in 1910 to more than 20,000 in 1941.[35]

In addition to quantitative demographic growth, Côtes-des-Neiges established its new function as a transnational migrant reception area during the interwar period. By World War II, Côtes-des-Neiges had become the center of Montreal's vibrant Jewish community. After the war, other migrants from Europe followed. And during the 1970s and 1980s, migrants came from every corner of the globe.[36]

By the 1990s, Côtes-des-Neiges and neighboring Nôtre-Dame de Grâce were home to more than 154,000 residents. This population made the area the most populous among the twenty-seven boroughs (*arrondissements*) created by a municipal reorganization that established Montreal's first ultimately unsuccessful unified metropolitan government in 2002.[37] Côtes-des-Neiges and Nôtre-Dame de Grâce housed a larger number of transnational migrants than any other borough at that time, trailing only Saint-Laurent in the percentage of its population that had been born outside Canada (46 percent for Saint-Laurent vs. 44 percent for Côtes-des-Neiges / Nôtre-Dame de Grâce).[38] About 60 percent of the neighborhoods' residents claimed to be able to speak French in 1991; 80 percent reported being able to speak English.[39] A larger percentage of the district's transnational migrants were "visible minorities" than anywhere else in Montreal. Such characteristics reinforced a popular perception of Côtes-des-Neiges and Nôtre-Dame de Grâce as an "immigrant area."[40]

Neighborhoods like Côtes-des-Neiges and Nôtre-Dame de Grâce nourished flourishing small business districts dominated by ethnic entrepreneurs.[41] By 1991, local residents were served by two dozen ethnic and neighborhood newspapers in various African languages, Arabic, Cambodian, Hebrew, Filipino languages, Lao, Vietnamese, Chinese, Latin American languages, Creole French, Québéois French, and English.[42] These papers counterbalanced the coverage of Côtes-des-Neiges and Nôtre-Dame de Grâce in the mainstream media, which, as with the Bronx, often focused on criminality and poverty.[43]

Physically located on the western slopes of Mont-Royal, and hence removed from downtown Montreal, both Côtes-des-Neiges and Nôtre-Dame de Grâce straddle traditional linguistic and socioeconomic fault lines.[44] Wealthy anglophones live nearby, while poorer francophones are to be found to the north and east. Interestingly, both well-off and poor households share the area as well, amplifying a promiscuous mingling of difference that so defines the area. The novelist Stephen Henighan effectively conveys the feeling of entering the area through the thoughts of an arriving Brazilian immigrant, who will emerge as a central character in *The Streets of Winter:*

> They turned into a residential area in which he detected a shadow of France—walls of sagging brick, each gabled house fused to its neigh-

bour—but as the taxi crossed a main street the French on the signs was interspersed with Greek lettering, a knotted Cyrillic chain, even a dash of Portuguese.[45]

Such neighborhoods as Côtes-des-Neiges and Nôtre-Dame de Grâce have not obliterated traditional anglophone West End / francophone East End residential patterns so much as they have muted them by creating a middle ground. Anglophone migrants from the West Indies, for example, continue to settle in neighborhoods that have long been home to migrants of African heritage from Nova Scotia, Ontario, and the United States; while Haitians seek out francophone neighbors.[46] Annick Germain and Damaris Rose captured this new metropolitan-scale ethnospatial reality in 2000, when they wrote:

> Recent studies carried out at a microscale, however, have not only un-covered the internal diversity of immigrant districts but have shown that the broad anglophone-francophone spatial divide has long been inter-sected and underlain by social and ethnocultural mix at a fine scale. The spatial patterning has, moreover, never been static and with the decline of Montreal's anglophone population and the diversification of immi-gration, forces are at work which, on the whole, point in the direction of increased ethnocultural heterogeneity throughout most of the Island of Montreal and in some of the off-island suburbs.[47]

Language policy proved to be perhaps as important in reshaping the me-tropolis as the changing complexion of newly arriving migrants from abroad. These highly controversial policies made a migrant's knowledge of French essential to his or her assimilation and socioeconomic advancement.

Growing concern over what francophones came to see as the fragile sta-tus of French in Quebec, in Canada, and in North America invigorated the drive for Quebec sovereignty among the province's electorate.[48] Succes-sive profederalist Liberal Party and sovereigntist Parti Québécois (PQ) provincial governments proposed increasingly stringent laws governing the use of language in public discourse. Quebec moved inexorably toward a so-ciety in which French was the most acceptable language for public speech, even as the Canadian federal government in Ottawa was promoting French–English bilingualism throughout the country.[49] These processes culminated with the passage of Bill 101 by Quebec's National Assembly in 1977.[50]

Bill 101 became a pivotal point in the initial legislative program of the PQ government led by René Lévesque, who had come to power for the first time in hotly contested elections in November 1976.[51] The language legislation, which was developed under the leadership of the minister of state for cultural development, Camille Laurin, sought to create an environment in which the French language would become the defining characteristic of Quebec life.[52] Bill 101 attempted to achieve this goal by regulating the use of the French language and other languages on public signs, at the workplace, and in educational institutions. The bill's most important provisions for this particular discussion were legally binding restrictions on school choice. Following the enactment of Bill 101, all children without two parents who had attended English-language schools in Canada (in other words, immigrants) were required to attend French-language schools.[53]

Within a generation, the Bill 101's educational requirements had become part of the rules of the game in Quebec life, welcomed by some and still bitterly opposed by others. Deidra Meintel and her colleagues, who were associated with the Montreal-based Groupe de Recherche Ethnicité et Société, concluded after examining language-use patterns among Montreal's immigrants two decades later that the city had evolved from a bilingual culture to an increasingly trilingual environment.[54] Immigrant children attended French-language schools and, simultaneously, often retained their heritage language. Therefore, Montreal's immigrants moved freely between a francophone environment at school or work and an allophone one at home. In addition, ambitious immigrant parents frequently ensured that their offspring became fluent in English so as to function successfully in the larger North American context. Over the course of a quarter-century following the enactment of Bill 101, Meintel argued, these patterns have nurtured an ethnic pluralism in which increasing numbers of Montrealers have moved freely among three languages in everyday life—switching back and forth among French, English, and their traditional language.

Transnational migrants thus came to contribute to the sustenance of Quebec's and Montreal's French character as more and more migrant children were being educated in French.[55] They did so, however, while simultaneously converting a French-language community into an intercultural society. As Dominique Arel argues, language use patterns in Quebec had changed fundamentally.[56]

"The Québécois in the cultural national sense," Arel writes, " are becoming less Catholic and also less white, but share a common identity by virtue of the fact that they speak French as a first language, the same way that the Canadians outside Quebec are becoming less Protestant and less white, but continue to adopt English as a mother tongue intergenerationally."[57] Modernization and declining population shifted the nature of discourse and rules of engagement between Montreal's francophones and anglophones. As noted above, French-speaking immigrants and their French-schooled offspring have become incorporated by state-imposed limits on educational choice into a general francophonic linguistic community that has protected the core of a French-language island in the English-language sea of North America.

If, as Arel has argued, cultural insecurity is a potent factor in political instability, this shift in the francophone perception of allophones as potential linguistic allies slowly diffused Montreal's contentious political life.[58] Immigrants were, in effect, no longer caught in the middle when they voted to undermine the secessionist endeavor, as they had in October 1995. The terms of discourse on language, sovereignty, and migrants had changed; as had residential patterns, which were becoming increasingly integrated (though at a slow pace). Arel has also suggested that "the secessionist movement will inevitably decline" because there is a "collective realization that immigrants are routinely becoming francophones," thereby making "the secessionist project less and less urgent."[59] This perception is especially true if present demographic trends hold constant, with allophones projected as overtaking the number of anglophones living in the Montreal region in the years ahead.

Montreal entered the twenty-first century with an astonishing mixture of ethnic, religious, racial, and linguistic communities—often without well-defined boundaries among them. Given the historic enmities between anglophones and francophones—as well as between Roman Catholic and Protestant communities—"it is," in the words of the *La Presse* journalist Laura-Julie Perreault, "something of a miracle that Montreal did not become a second Northern Ireland."[60] Instead, as Perreault reports, the city's 350-year-old multiconfessional regime has become a marketing tool for its burgeoning tourism industry.

One particularly curious story captures the tenor of twenty-first-century Montreal's melange of overlapping cultures and religions. In February 2004, Filipino and Hispanic parents started to bring their sick children to a small apartment complex in Saint Laurent to be cured by an icon-like portrait of the Virgin Mary that had been discovered in the garbage by a Muslim apartment building manager, Abderezak Mehdi. The portrait—according to Mehdi and a Greek Melkite Catholic priest, Michel Saydé—shed tears of oil that could cure the ill. Meanwhile, Michel Parent, the chancellor of the Roman Catholic archdiocese of Montreal, cautioned skepticism, noting that "while it is true that nothing is impossible for God, historically, that is not how God acts."[61]

New patterns of linguistic, ethnic, social, and cultural interaction may also be seen in how Montrealers have come to use public space. Seguin and Germain contend that, by the beginning of the twenty-first century, systematic observation of the city's public spaces reveals "a common pattern that could be described as a peaceful but distant cohabitation of users of diverse ethnic origins. In these spaces, residents seem to respect a code of civility, which enables them to enjoy the diversity of social contact offered within these spaces, while maintaining distance from other users."[62]

The result, as one can see in Stephen Henighan's popular novels of late-twentieth-century Montreal, becomes a city in which, to quote from a review of Henighan's 2004 story *The Streets of Winter*, residents "are unable, or unwilling, to relinquish the individual and cultural prejudices, nostalgias and expectations that alienate them in their quests for 'le vrai Montréal.'" Once starkly divided by language, Montreal has become converted into "a grid of many solitudes."[63]

Haunted by the ghosts of the past though a search for "the authentic Montreal" may be, the city's contemporary embrace of cultural, religious, ethnic, and linguistic difference is markedly more enthusiastic than at any other moment in its troubled recent history. Montreal's diversity capital increased dramatically over the course of the late twentieth century. Hugh MacLennan's "two solitudes" have indeed been replaced by "multiple solitudes."[64]

## Washington: Fluid Racial Lines

The greater Washington-Baltimore metropolitan region entered the twenty-first century as the nation's fourth largest urban region, one that was 25 percent black, 69 percent white, and 6 percent other; as opposed to the District of Columbia, which then was 60 percent black, 31 percent white, and 9 percent other.[65] These figures suggest that Washington and its region have remained racially segregated at the aggregate level despite significant changes in national and local legislation.

A much more variegated pattern emerges at the micro level of neighborhood and community. The removal of laws and customs restricting where African-Americans could live within the District of Columbia a half-century ago, together with the ending of school segregation, led to dramatic "white flight" across state lines into neighboring Maryland and Virginia. According to U.S. census data, Washington's white population declined by 308,593 between 1950 and 1970.[66] As the data in table 2.1 suggest, the city's white population subsequently stabilized during the 1980s.

Washington's black population increased from 280,803 in 1950 to 527,705 in 1970. Its robust African-American middle class then began its own exodus to the suburbs (primarily to Prince George's County, Maryland). By 2000, Washington's black population had declined to 346,382. Meanwhile, the number of city residents who are neither black nor white has grown exponentially, from just 3,510 in 1950 to 9,526 in 1970 and 63,518 in 2000. Not surprisingly, such dramatic demographic and social upheavals have affected nearly every neighborhood and street corner in the city.

City planners contributed to Washington's social disruptions during these years as well. As in many U.S. cities, planners attempted to make the city more convenient for automobiles through an ambitious highway construction program. The now-famous Washington Beltway was planned, together with freeways from suburban outskirts into the city center linked by an inner highway loop embracing the federal center downtown.[67] As was so often the case with such proposals, local administrators intended many of these new roads to run through the poorest African-American sections of town. The goal, in the memorable words of architectural critic Wolf Von Eckardt, was to bring a combination of "suburban wholesomeness with urban stimulation" to downtown.[68] Official Washington would accomplish this lofty end by removing African-Americans from view.

Led by the clarion voice of the *Washington Post,* social reformers and city planners attacked the old African-American neighborhoods of the city's

*Table 2.1. District of Columbia Population, by Census Category, 1970–2000*

| Year | Total | White | Black | Latino | Other |
|------|---------|---------|---------|--------|--------|
| 1970 | 756,668 | 210,863 | 527,705 | 15,617 | — |
| 1980 | 638,333 | 164,244 | 445,154 | 7,679 | 11,256 |
| 1990 | 606,900 | 166,131 | 395,213 | 32,710 | 12,864 |
| 2000 | 572,051 | 162,121 | 346,382 | 44,953 | 18,595 |

*Source:* Krishna Roy, "Socio-Demographic Profile," in *The State of Latinos in the District of Columbia,* ed. Council on Latino Agencies / Consejo de Agencias Latinas (Washington, D.C.: Council on Latino Agencies / Consejo de Agencias Latinas, 2002), 1–30 (fig. 1.1, p. 5); available at http://www.consejo.org.

Southwest quadrant throughout the 1950s. "No doubt many residents of the area will be loath to lose their homes despite the prevailing slum conditions," the *Post*'s editorial writers would observe. "They should realize, however, that the net effect of this great redevelopment effort will be to make Washington a much more pleasant place in which to live and work."[69]

The consequent fight against highway construction was similar to, yet different from, what transpired in other U.S. cities at the time because Washington is unique among U.S. cities in having no legal jurisdiction over itself. The city was governed for much of the twentieth century by commissioners appointed by the federal government. Defeating various projects, such as the inner Beltway and the Three Sisters Bridge across the Potomac River just north of Georgetown, required that local activism and street demonstrations be carried out in conjunction with carefully orchestrated elite negotiation among local notables, senior White House officials, and congressional figures.[70] Local governance could not be separated from national politics in a city so long denied even a modicum of home rule.

Washington finally gained limited local autonomy to govern its own affairs only in January 1975, in accordance with the Home Rule Bill that had been signed by President Richard Nixon on Christmas Eve 1973 (though Congress retained final approval authority over all D.C. legislation and budgets).[71] The movement behind local rule was mobilized in part by resistance to the plans of federal overseers to bulldoze a ring of highways through the heart of several intown neighborhoods. As always in Washington, such intensely local disputes over urban form could never be distinguished from the larger politics of race.

The highway battles' denouement ended with the destruction of poor African-American neighborhoods in Southwest.[72] Local activists gamely fought back the bulldozers in other areas—such as the Shaw district in Northwest (though they could not stave off more prosaic forms of urban decline and rising crime). Still other parts of the city—such as the West End— subsequently became gentrified, with young professionals moving in once the threat of a new freeway had been lifted. The wealthy and predominantly white enclave of Georgetown eventually fought off the intrusion of freeway approaches to a new major bridge across the Potomac.

This general reshuffling of inner-city neighborhoods had run its course as transnational migrants began to come to the Washington area in visible numbers. Where highways were ultimately built or not, combined with the simultaneous lifting of restricted covenants prohibiting nonwhite ownership of homes and an end to school segregation, all locked in place new spatial patterns that would define the early-twenty-first century Washington cityscape.

The Southwest section of the city, for example, has become the sort of urban neighborhood that 1950s planners so desperately wanted to build. Knock-off Corbusierian towers are scattered without reference to traditional street plans, with no poor neighborhoods left to upset the tourists as they sit in monumental traffic jams on the Southeast-Southwest Freeway. Of the 22,539 residents displaced from this project, 80 percent were African-Americans and nearly all were poor. About a third would find alternative homes in public housing, while 2,000 families moved into private rental units, and only 391 families were able to purchase private homes in other parts of the city.[73]

As these numbers imply, poor black folk from Southwest moved into other neighborhoods around town, overloading formal and informal social networks and support systems that had developed over decades. Moderately prosperous African-Americans, in turn, left their older neighborhoods for the more comfortable houses being abandoned by whites fleeing for the suburbs. The 1968 riots following the assassination of Martin Luther King Jr. finalized this process by destroying several previously vigorous African-American shopping districts. Once the dust had settled on the 1960s, Washington's African-American community was arguably more segregated and isolated than when the decade had begun. A hard racial frontier traced Rock Creek Park from the city's northern border with Silver Spring, Maryland, down along Georgetown's edge.[74]

As the census data mentioned above reveal, Washington hit a low point in its history as an urban community (rather than as a symbolic capital city) sometime between the 1968 riots following the assassination of Martin Luther King Jr. and the introduction of home rule in 1975. The U.S. bicentennial celebrations, the opening of the Metrorail subway system, home rule, the appearance of dynamic "new generation" political leadership in the person of D.C. Council member Marion Barry—such events pointed to a bright future for the city. David L. Lewis captured this mood in the closing to his history of the city published that year: "Black or white, rich or poor (though most will be richer than poorer), heavily taxed, denied political equality, [Washingtonians] will continue to grow into the comity of their new roles of cultural pace-setters and urban innovators."[75] This was precisely the moment when migrants from all over the globe began to arrive in the city for the first time in numbers large enough for their new neighbors to notice.

Two trends have combined to redefine the city's social landscape during the subsequent quarter-century. First, an increasingly wealthy African-American professional and middle class began its own exodus for the suburbs—primarily to Prince George's County east of Washington. As noted above, the city's black population declined by nearly 200,000 between the 1970 and 2000 censuses. By 2000, nearly as many blacks were counted by census takers in Prince George's County as there had been people of all races residing in that jurisdiction in 1970 (about half a million residents).[76]

Thus, at the aggregate metropolitan level, the historic large-scale segregation of the region's African-American population continues, because blacks most frequently reside in the once plantation-dominated coastal plain south and east of the Potomac fall line (see table 2.2). Within the District of Columbia, those wards with above-average concentrations of foreign-born and white residents similarly are to be found north and west of the fall line, with those neighborhoods housing above-average percentages of African-Americans located to the south and east.[77] This large-scale pattern persists even though micro-level neighborhood and community analyses reveal heightened racial and ethnic integration throughout the region.

Second, the city's transnational migrant population increased nearly fivefold, as noted above. The proportion of the overall population born abroad rose from 4.4 percent in 1970 to 9.7 percent in 1990, and nearly 13 percent in 2000.[78] The city's new arrivals came largely from Central and South America, together with significant numbers of migrants from the Caribbean, Southeast Asia, and Africa.[79]

These new Washingtonians have settled primarily in such central-city neighborhoods as Adams Morgan and Mount Pleasant, as well as in the adjoining areas of Columbia Heights and Cardozo-Shaw.[80] Adams Morgan and Mount Pleasant are, in many ways, Washington's equivalents to Montreal's Côtes-des-Neiges and Nôtre-Dame de Grâce.[81] In the assessment of the Washington Council of Latino Agencies, both neighborhoods sustain "a strong multicultural community identity through a network of multiracial and multilingual community-driven services and activities."[82]

The city's Latino residents in particular have concentrated in central Ward One areas along the once-impenetrable racial boundary of Sixteenth Street, Northwest. In 2000, Ward One's census tract 28.2 reported the largest percentage of Latino residents among all the city's census units (51 percent).[83] This vibrant migrant community has come under increasing pressure from rising real estate and rent prices and an accompanying process of rampant gentrification.[84] As a result, the 2000 census reported

Table 2.2. *Percentage of Population Identified as*
*"Black" in Metropolitan Washington, by County, 2000*

| Jurisdiction | Percentage Black |
|---|---|
| District of Columbia | 60.01 |
| Maryland | |
|     Anne Arundel County | 13.57 |
|     Calvert County | 13.11 |
|     Charles County | 26.06 |
|     Howard County | 14.42 |
|     Montgomery County | 15.14 |
|     Prince George's County | 62.60 |
|     Saint Mary's County | 13.92 |
| Virginia | |
|     Arlington County | 9.34 |
|     Alexandria | 22.54 |
|     Fairfax County | 8.57 |
|     Fauquier County | 8.78 |
|     Frederick County | 2.62 |
|     Loudoun County | 6.89 |
|     Prince William County | 18.76 |
|     Stafford County | 12.12 |

*Source:* U.S. Census Web site: http://factfinder.census.gov.

notable increases in the Latino populations of Wards Four, Five, and Six as well.[85]

Washington's transnational migrants have come to fill an intermediary socioeconomic space. Robert Manning reported, for example, that Hispanics earned a median household income of $26,000 in 1989, while Asian city residents were earning $30,000. These figures were slightly higher than the median household incomes of Washington's African-Americans at the time ($25,000); and all these groups remained significantly behind local whites in earning power (the median household income of the city's whites was reported in the 1990 census as $46,000).[86]

This hierarchy is further evident in the child poverty rates among various groups within the city. For example, the poverty rate for the city's African-American children in 2000 was 38 percent, as opposed to 26 percent for Latino children. A total of 89 percent of the city's children living in poverty were African-American, in contrast to 8 percent of the city's poor children being of Latino origin. The figures for other groups are negligible in comparison.[87]

Thus, in the last third of the twentieth century, Washington emerged as a major center of transnational migration for the first time in its history. As important, the city stood at the center of a sprawling metropolitan region where migrants from abroad were creating a middle ground beyond the traditionally racially divided center city. During the 1970s and 1980s, more than seven times the number of immigrants residing in the city were living in the much larger Washington metropolitan region (see table 2.3).[88]

Like their native-born neighbors, transnational migrants now live farther and farther out within the Washington metropolitan region. The 2000 census reports that the largest growth rate for suburban Latino communities during the 1990s was to be found in distant Loudoun County, Virginia (with a 37 percent growth rate), and Frederick, Maryland (with a 17 percent growth rate).[89] By 2000, immigrants in the city were more likely to be poor than immigrants elsewhere in the region (though less likely to be poor than the city's native-born population) and less likely to speak English.[90] This metropolitanization of transnational migrant life reflects a larger pattern throughout the United States during the 1990s, as migrants from abroad were finding their way to suburbs with previously unknown frequency.[91]

The scale and extent of late-twentieth-century transnational migration to Washington and its surrounding metropolitan region have been most ex-

Table 2.3. Percentage of Population Foreign Born in Metropolitan Washington, by County, 2000

| Jurisdiction | Percentage Foreign Born |
| --- | --- |
| District of Columbia | 12.86 |
| Maryland | |
|    Anne Arundel County | 4.74 |
|    Calvert County | 2.20 |
|    Charles County | 2.88 |
|    Howard County | 11.34 |
|    Montgomery County | 26.68 |
|    Prince George's County | 13.78 |
|    Saint Mary's County | 2.82 |
| Virginia | |
|    Arlington County | 27.81 |
|    Alexandria | 25.41 |
|    Fairfax County | 24.51 |
|    Fauquier County | 3.59 |
|    Frederick County | 3.98 |
|    Loudoun County | 11.27 |
|    Prince William County | 11.46 |
|    Stafford County | 4.02 |

Sources: U.S. Census Web site, http://factfinder.census.gov; "Washington by the Numbers: Our Region, According to the 2000 Census," Washington Post, December 8, 2002.

tensively analyzed by Audrey Singer and her colleagues in a study for the Brookings Institution. Singer and her colleagues set out new terms for considering Washington as one of the country's most dynamic metropolitan regions.[92] Working with data from the Immigration and Naturalization Service linked to postal zip codes from 1990 to 1998, rather than with census data from 1990 and 2000, these researchers captured the profound impact that Washington's new transnational migrants are having on neighborhoods and communities throughout the region. They summarized their major findings as follows:

—In 1998 the Washington metropolitan area was the most common destination for legal immigrants. Only New York, Los Angeles, Chicago, and Miami were more popular. Between 1990 and 1998, nearly 250,000 immigrants from 193 countries and territories chose to live in the metropolitan area.

—Washington's recent immigrants are highly diverse—there is no dominant country (or countries) of origin among the newcomers to the region. The largest single immigrant group—from El Salvador—comprises only 10.5 percent of the region's newcomers.

—Washington's immigrants are not clustered into ethnically homogeneous residential enclaves, but instead are dispersed throughout the region. Of the top ten immigrant zip codes, four each are located in Maryland and Virginia, and two are in the District of Columbia.

—In the 1990s, 87 percent of immigrants to the region chose to live in the suburbs. Almost half (46 percent) of new immigrants located in communities outside the Capital Beltway. Less than 13 percent moved to the District.

—Asian immigrants are more likely to move to the outer suburbs, while Latin American and African immigrants tend to live within the Beltway.[93]

The diversity of metropolitan Washington's transnational migrant population was reflected in the fact that several zip codes in the region had become home to "newcomers from over 130 countries."[94] Singer and her colleagues reported further that 42.0 percent of the region's migrants came from Asia (including 7.4 percent from Vietnam, 6.6 percent from India, 4.6 percent from China, 4.4 percent from the Philippines, and 4.1 percent from South Korea), 31.5 percent traveled from Latin America and the Caribbean (including 10.5 percent from El Salvador, 2.9 percent from Peru, 2.3 percent from Bolivia, and 2.1 percent from Jamaica), 16.2 percent arrived from

Africa (including 3.9 percent from Ethiopia, 2.3 percent from Nigeria, and 2.0 percent from Ghana), while only 10.3 percent originated in countries of Europe, Oceania, and Canada (including 2.7 percent from the countries of the former Soviet Union and 1.6 percent from the United Kingdom).[95]

Today's Washington metropolitan region is dramatically different from the "Chocolate City, Vanilla Suburbs" of 1970. As Singer and her colleagues demonstrated, neighborhoods across the entire metropolitan region had been transformed in a generation into some of the most ethnically diverse communities in the United States. Zip code 22204 in South Arlington, Virginia, as well as zip code 20009 in Adams Morgan–Mount Pleasant, Washington, were among the major transnational migrant urban communities in the United States.[96] New diversity capital is being created every day as longtime residents and new arrivals learn to accommodate one another. In doing so, they slowly create the building blocks for "policies and institutions that have the overall effect of integrating diverse groups and cultural practices in a just and equitable fashion," which can become the essence for societal capacity to absorb diversity.[97]

Race has hardly disappeared as a major dimension in metropolitan Washington life. Many of the region's most diverse neighborhoods—such as Adams Morgan and Mount Pleasant—sit astride historic Washington racial dividing lines such as Sixteenth Street, Northwest. African migrants are most likely to reside in Washington and Prince George's County.[98] Moreover, migrants of color from Africa and Latin America predominately settle inside the Washington Beltway, a ring road that demarcates an increasingly sharp cultural boundary between "urban" and "suburban" lifestyles (i.e., people still walk from time to time inside the Beltway, but residents of the region beyond the Beltway are dependent on the automobile).[99]

The *Washington Post* chronicled these changes in a series of articles during 2002. The paper's reporters discovered a new Hispanic presence in once predominantly African-American neighborhoods such as Petworth along Fourteenth Street, Northwest.[100] They recorded the surprise of some darker-skinned migrants from Latin America at being considered "black" for the first time in their lives.[101] As tellingly, they note "a little-known reality: America's black community, which now includes more West Indian and African immigrants than ever, is no longer the monolithic group that many politicians, civil rights advocates, and demographers say it is."[102] Darryl Fears, the *Post* reporter who wrote many of these pieces, concluded that "a new African-American community is being forged . . . in which culture and nationality are becoming more important than skin color."[103]

As in Montreal, three decades of expanding migration from every corner of the globe—except from Europe—is transforming where and how Washingtonians live. The Washington region has become metropolitan in scale and, in the process, has evolved further along the path from a biracial to an intercultural community than many residents and politicians recognize. Race still matters, to be sure. For the first time, however, Washington's blacks can be "ethnic" as well. A new transnational diversity is very slowly beginning to mute old racial hostilities. The city's store of diversity capital has increased.

## Kyiv: A New Transnational City?

Kyiv attracted its fair share of Soviet-era migrants, especially peasants moving to town from the countryside.[104] These migrants were largely ethnically and linguistically Ukrainian. For example, the demographer Irina Pribytkova has noted that more than half the city's residents at the time of the last Soviet census in 1989 had been born elsewhere.[105] One-fifth of the Kyivans at the time had lived in the city for fewer than five years, and another third had moved to town more than a quarter-century before. The majority of these new Kyivans were transplants from collectivized villages in Ukraine, with nearly another third having moved from other large Ukrainian cities. One nonnative Kyivan in ten came from Russia, being joined in the city by a scattering of migrants from other Soviet republics.[106] Thousands more Soviet-era migrants came to the city as so-called *limitchiki* to work at local bread- and meat-processing plants and railroad-car factories. These new arrivals were permitted to forgo the rigors of the Soviet residential *propiska* (permit) system, as long as they worked in labor-starved industries.[107]

Ukraine's initial postindependence years saw a final sorting out of these Soviet-era migratory patterns—through family reunification, the repatriation of Ukrainians from other Soviet republics, and the relocation of residents from areas contaminated during the Chernobyl' nuclear reactor accident.[108] The city's official population rose to an all-time high of 2,626,500 in 1994, when the post-Soviet reshuffling of Ukrainians and Kyivans was largely complete.[109] By the late 1990s, Kyiv's traditional migratory pool had all but run dry, because the city's stumbling economy no longer produced the jobs that would have attracted new workers. The local population began to decline as a consequence of both net natural population loss (deaths surpassing births), and modest out-migration back to the countryside.[110] In addition, an unrecorded and probably significant number of residents migrated abroad.[111] It was about this time—the mid-1990s—that transnational migrants began to appear in the Ukrainian capital in significant numbers.[112]

Kyiv's postindependence migrants generally fall into one of three groups. First, migrants from formerly socialist countries and countries sympathetic to the Soviet Union arrived as students or guest workers of various sorts at the end of the Soviet period and remained. Second, refugees (both officially recognized and not) fled conflicts in other regions of the former Soviet Union, the Balkans, and the Middle East to settle in Kyiv. Third, "irregular migrants" of various sorts made their way into Ukraine, often in search of an easy route into Europe. Illegal trafficking in migrants has grown during the past decade as Ukraine has become an alternative route into Europe to migration circuits disrupted by the Balkan wars of the 1990s.[113]

Motivations for traveling to Ukraine and modes of travel have varied both across and within these three groups. Migrants themselves most often cite economic considerations, such as low standard of living, difficulties in finding employment, and limited opportunity for financial gain in their native lands.[114] In addition, about one-quarter of arriving migrants in Ukraine at the end of the 1990s identified wars, armed conflicts, and political instability at home as the primary causes for their decisions to migrate. Ukraine became a destination of choice due to its perceived low cost of living, supposedly easy access to Western visas, and the presence of friends and relatives already living there.

Migrants have followed a variety of routes to the Ukrainian capital, sometimes overstaying legal visas, and other times purchasing illegal "packages" from criminal "travel agents" abroad or arriving with the help of forged documentation and bribes. Still others merely rushed through seldom-patrolled Black Sea ports and Ukraine's largely unguarded frontier with Russia.[115] Whatever their initial intent, and however they arrived, life's normal ebbs and flows conspired to encourage tens of thousands of individual migrants to settle in Kyiv and to build new lives for themselves in Ukraine. Increasingly, they do so as citizens of Ukraine.[116]

Once they arrived in Kyiv, migrants began settling in apartment buildings that were coming to be viewed increasingly as less desirable by indigenous Kyivans. A new housing market—rather than state-controlled administrative agencies—redistributed apartments in the city. Though not literally places "in between" linguistic or racial communities, as was the case in Montreal and Washington, these neighborhoods were metaphorically "in between" the old Soviet planned economy of the past and a new real estate market that was coming into being following Ukraine's independence. In this sense, new migrants were once again creating a new middle ground.

Three outlying Soviet-era Kyiv neighborhoods have emerged as migrant centers. Troeshchyna to the east and Syrets to the north are especially noteworthy as major multiethnic centers in the making, with perhaps as many as half the city's migrants living in these two shoddy fringe districts. Borshagovka, to the west, has become a focal point for the city's small but growing African community. No other neighborhood accounts for a significant concentration of migrants, while a handful of more prosperous intown neighborhoods such as Pechersk and Podil' have been unaffected by the new migrants in any meaningful way.[117]

Troeshchyna, in particular, is emerging as a classic migrant neighborhood, becoming home to several groups with relatively self-contained lives tied to a burgeoning informal market.[118] It is, in many ways, Kyiv's new neighborhood like Montreal's Côtes-des-Neiges and Nôtre-Dame de Grâce or Washington's Adams Morgan and Mount-Pleasant—but dressed in distinctive post-Soviet garb. Troeshchyna's housing stock is a product of official policies and practices of the late Soviet period;[119] the neighborhood was built by government-run construction firms and operated by municipal offices and state enterprises, after first being identified for development in the 1967 Kyiv General Plan.[120] Planners and architects announced more detailed plans for Troeshchyna in 1981, projecting the emergence of a new district of the "highest sanitary-hygienic conditions and comfort" housing 300,000 residents.[121]

Widespread praise for the architects' plans followed in both the popular and professional press, as Troeshchyna was held up as an example of the loftiest achievements of Soviet city planning.[122] Local authorities were able to have Troeshchyna designated an "experimental" district, which granted them greater latitude in design and construction standards. The area's cheesy prefabricated colored cement panels and tiled portrayals of Ukrainian folk motifs were the result of efforts by local architects to use this additional license to "humanize" the area.[123] The neighborhood's built environment was proclaimed to be the pinnacle of official Soviet housing and design policies, as indicated by its designation as one of a small handful of "experimental" city planning projects across the entire Soviet Union.

Migrants turned to this area after the creation of a large informal market on vacant land within the Troeshchyna neighborhood. This market emerged in April 1996, when officials concerned with the city's image—and encouraged by many disgruntled Kyiv residents—drove traders from Kyiv's largest sporting venue, the Republican Stadium, where they had taken up business during the waning months of the Soviet regime. In scenes reminiscent of similar confrontations in many poor cities around the world, local police raided the market and physically forced the traders to move elsewhere following a campaign by the mayor and other city officials in the name of public safety and security, and the well-being of local sports.

The indigenous traders dislodged from the Republican Stadium market were able in some instances to draw on previous connections to ensconce themselves in new stores and stalls at officially sanctioned locations elsewhere in the city center. Those selling cars and cigarettes, for example, drew on underground contacts with local politicians to find "civilized" outlets in equally lucrative locations around the city.[124]

The migrant merchants, being foreign, were forced to leave downtown because they had fewer opportunities to seek political sponsorship from above. Several traders—including a contingent of Afghan migrants—converted the polluted vacant fields of a sprawling abandoned defense plant along the city's northeastern boundary at Troeshchyna into a boundless tent city.[125] Once beyond official control and concern, the merchants set about bringing a rough-and-tumble order to their haphazard market. Passages through the stalls became streets specializing in a particular category of goods—furniture here, clothes there; household items down one alley, videos and compact discs down another. Trading customs and business norms emerged, often enforced by the sort of unofficial paramilitary groups and vigilante gangs known throughout the former Soviet Union as "the mafia." Within two years, more than 5,000 stalls at the Troeshchyna market employed some 20,000 Kyivans.[126]

According to recent surveys, Troeshchyna migrant residents have twice as many contacts with their own migrant compatriots at work, at home, and in leisure activities than with native Kyivans.[127] Troeshchyna's migrant residents also report low levels of Russian and Ukrainian use and knowledge, more frequently conversing with friends and family in their native languages.[128] In short, the district's immigrant communities increasingly function apart from the rest of the city.

Though less dramatic than Troeshchyna, many similar patterns are beginning to manifest themselves in Syrets to the north. As in Troeshchyna, Syrets appears to be an immigrant district in formation, with residents living in two self-contained communities: one of transnational migrants, and one of native Kyivans.

Borshagovka, home to the vast majority of African-Kyivans, is a small world in which residents regularly encounter the city at large rather than remaining in a closed circle of their fellow migrants. The knowledge and use of Russian and Ukrainian among Borshagovka migrants is high, while Borshagovka transnational residents more often come into contact with the native Kyiv population at work and at home.[129]

Perhaps the most striking aspect of Troeshchyna, Syrets, and Borshagovka is that they are products of Soviet-style metropolitan expansion. All three lie well beyond Kyiv's mid-twentieth-century boundaries, representing a late-Soviet-era neighborhood type, one that was tied to large-scale regional settlement and land-use patterns.

Transnational migrants in Kyiv—as in Montreal and Washington—arrived just as the city expanded into a metropolitan-scale region. Migrants similarly turned these Soviet suburbs into intercultural neighborhoods, sites for the creation of a postindependence pragmatism concerning diversity. Though the groups conduct their daily lives largely in relation to one another, they live side by side with the Russians and Ukrainians who came to these districts during the Soviet era. If diversity capital along the lines of Stren and Polese's "urban social sustainability" is to emerge in Kyiv, it will be in these outlying Soviet mega-neighborhoods.[130]

New migrants from abroad are turning Kyiv into an intercultural, bilingual metropolis. Language remains Kyiv's most evident defining individual characteristic. Migrants learn to function in both Russian and Ukrainian, an adaptation that is necessary as Ukrainian has become the only language used in government-related transactions in an independent Ukraine.[131] Meanwhile, Russian floats through the air, most noticeably at the Troeshchyna market.

Migrant children attend local Russian and Ukrainian language schools, often becoming trilingual in the process. As in Montreal and in Washington, their linguistic ambiguity mutes long-standing divisions within the city. In a very real sense, transnational migrants are converting Kyiv into an intercultural, multilingual metropolis.

Kyiv, like Montreal and Washington, has evolved in recent years from a centralized, linguistically bifurcated city into an intercultural, multinodal, metropolitan region. In fact, the reshaping of urban physical space may be somewhat further along in Kyiv than in Montreal and Washington, because Soviet state-dominated city-building efforts were never as amenable to group self-segregation as those of North America's private real estate markets. Identity conflicts were—and remain—largely interior to individual Kyivans rather than having been etched on city maps.

Rapid metropolitanization during the late twentieth century in Kyiv, as well as in Montreal and Washington, has reshuffled residential patterns. Large-scale suburban development—albeit in multiapartment high-rise towers rather than single-family houses—has converted residents into neighborhood newcomers as the rules, customs, and procedures for securing living space have changed.

The simultaneous arrival of migration from abroad and metropolitan-scale development have converted Montreal, Washington, and Kyiv from cities where, to borrow from Mordecai Richler, neighborhoods were a "sequence of ghettos," into sprawling intercultural metropolitan regions spreading out over hundreds of square miles and kilometers. This physical growth in scale and complexity has sustained new "policies and institutions that have the overall effect of integrating diverse groups and cultural practices in a just and equitable fashion" in all three cities. Metropolitan expansion is creating the possibility for the emergence of fresh stores of diversity capital in Montreal, Washington, and Kyiv alike.[132]

Older macro-level patterns of segregation persist at the aggregate level, at least in Montreal and in Washington. Neighborhoods east of Saint Lawrence Boulevard/Boulevard Saint-Laurent are still predominantly francophone, just as neighborhoods east of Sixteenth Street, Northwest, continue to be predominantly African-American. The intercultural transformation of residential patterns in all three metropolitan regions into a new middle ground becomes most visible at the micro, neighborhood level.

As dramatic as these changes have been in Montreal's, Washington's, and Kyiv's neighborhoods, the new intercultural identities prompted by the arrival of migrants from abroad have been even more pronounced at the workplace. As will be discussed in the next chapter, it is on the job and at school—rather than at home in their neighborhoods—that typical Montrealers, Washingtonians, and Kyivans encounter people of difference. The workplace and schoolroom may well hold more powerful clues about the extent to which all three cities have more resources at their disposal with which to accommodate diversity.

# Chapter 3

# Working and Studying in the Middle

Me and Yvette are drinking coffee and trying to find out about school.
Think about it. Is she white like me or brown like Yvette?
She's green, the kid says.
Yvette laughs.
I told you they was weirdos at that school. They let anyone in. They don't even
ask what planet you're from.

—Ursula Barnes, "Every Colour under the Sun," 2003[1]

While acknowledging the importance of residential patterns among various
ethnic, linguistic, and racial groups for the functioning of a diverse metropolitan region, urban observers should recognize that where and how people work and go to school may be even more critical to the success of intergroup accommodation in today's cities. Many members of self-identified
groups seek out those who are similar to themselves as neighbors. Even if
a metropolitan region is not divided into precise ethnic, linguistic, and racial
enclaves, many residents nonetheless may choose to live their personal lives
within the bounds of their own group. Salvadorans in Alexandria, Virginia,
may spend more of their personal time in contact with Salvadorans living
in the Mount Pleasant area of Washington than with their more immediate
Virginia neighbors. Where people live—and with whom they interact—is
largely (though hardly solely) decided by personal choice.

Richard Wright and Serin Houston, among many, have argued persua-
sively that daily existence at the workplace and in the classroom is significantly different from that at home and in the neighborhood.[2] The office,
store, and factory as well as the school and college are where the twenty-

first-century North American city's new middle ground is created. Once on the job or at school, denizens of diverse metropolitan communities must work with those whom they otherwise have had scant contact (or, quite possibly, do not view with affection). Personal choice becomes more limited when income depends on finding a way to perform a task in cooperation with others. The worlds of work and school now serve as perhaps the most important venues for the accumulation of new diversity capital.

Tensions ensue as the worlds of work and of school are hardly more harmonious than those of a neighborhood. Rather, work and school become the places where transnational migrants in cities like Montreal, Washington, and Kyiv are unavoidably tossed into the maelstrom of group interaction. Vietnamese migrants to Montreal must learn the delicate dance among those moments when it is obligatory to speak French and when it is acceptable to speak English; Vietnamese migrants to Washington must learn how to navigate between and among whites and blacks in a world supercharged by race; Vietnamese migrants in Kyiv must manage their official contacts with government offices in Ukrainian even as they negotiate at a market stall in Russian. Work and school environments are different from what is at home in that they remain places where migrants are often placed at a disadvantage.

In addition to the presence of newly arriving transnational migrants, Montreal, Washington, and Kyiv stand at the center of large metropolitan regional economies that have been undergoing rapid structural change. Montreal spent the late twentieth century confronting the perplexities of deindustrialization; Washington expanded its economy well beyond a once-dominant public sector; Kyiv both deindustrialized and shifted the focus of economic growth from the state to the private sector. As a result, transnational migrants in these particular metropolitan communities face ever-changing rules of local economic life even as they themselves were trying to master what to them may be alien rules of intergroup relations.

All three regions simultaneously have been transformed physically and spatially by massive expansion outward, moving employment and educational sites away from the center city to more amorphous neighborhoods on the edge of the city. Suburban sprawl contributes to a work world where everyone commutes to work and goes to school (from kindergarten through to graduate school) with others who have similarly left friends and family in neighborhoods far away. Montrealers, Washingtonians, and Kyivans regularly spend their workdays and schooldays with colleagues from different parts of the world who reside at opposite ends of the metropolis.

Migrants from abroad and natives meet others unlike themselves, creating some of the preconditions for "policies and institutions that have the over-all effect of integrating diverse groups and cultural practices in a just and equitable fashion," which can become the essence for societal capacity to absorb diversity.[3]

## Montreal: A "Daily Dance" of Language and Cultures

Nothing in Montreal easily transcends language. Just as school systems were divided by race in Washington (and in many other jurisdictions throughout the United States) until the second half of the twentieth century, Montreal's anglophone Protestants and francophone Roman Catholics or-ganized separate—and hardly equal—school systems in 1857.[4] Anglophone elites proved more generous with their schools than their francophone neighbors, supporting a hierarchy of premier educational institutions lead-ing up to and including McGill University. French Catholic educational in-vestment focused on reproducing pastoral Quebec society, with seminaries representing the pinnacle of francophone educational achievement.

This dual system was replaced at the end of the twentieth century by one organized around language—with two separate boards (the English Mon-treal School Board / Commission Scolaire English–Montréal; and the Com-mission Scolaire de Montréal) coming into existence. This reform of the tra-ditionally religious-based Montreal school system acknowledged various changes in education mandated by Bill 101, which became law in 1977.

As already noted, Bill 101 effectively ended the ability of anglophones and allophones to isolate themselves from the French fact of Montreal's daily life. Anglophones with no interest in learning French fled to Ontario and other destinations further west; transnational migrants often had scant choice but to learn French themselves and to send their children to fran-cophone schools. This new educational reality transcended neighborhood boundaries, tossing children into the wrangle of intercultural and multilin-gual education. Such a metamorphosis at school blended into an increasingly complex reality in the world of work. Following the enactment of Bill 101, language set down a new rhythm for what Elke Laur has called the "daily 'dance of life' of Montreal."[5] Montrealers of all ages could no longer seek refuge in tightly drawn, ethnically unified, monolingual neighborhoods.

The University of Montreal sociologist Christopher McAll has argued that his city's often noted residential frontiers have become both less stable and less critical for capturing the complexity of Montreal's multiethnic character than is commonly observed.[6] "Montreal's language frontier," McAll writes, "is frequently seen in residential terms, with the English-speaking population being heavily concentrated in areas west of the city centre. However, the highly visible (and audible) language frontiers between the different urban areas, along with the mixed-language frontier zones, are in many ways less significant in terms of the processes underlying the construction of boundaries than the more complex, less visible, but omnipresent frontiers that run through the Quebec workplace."[7] With transnational migrants constituting nearly a fifth of the Montreal metropolitan area's employed labor force throughout the 1990s, new Montrealers frequently found themselves caught in the middle of the city's complex language realities once they left home and entered the world of work.[8] As Wright and Houston observed for the United States, McAll argues convincingly for Canada that the world of work is central to an understanding of effective urban diversity management.

Montreal's workforce is highly segmented by profession, gender, nationality, and economic sector.[9] On one level, the city's employment patterns parallel those found in other major Canadian cities. Immigrants have higher unemployment rates and lower wages than Canadian-born workers, even when differences in education levels are taken into account.[10] Similarly, women—especially female transnational migrants—are disadvantaged in local labor markets.[11] Wages in the financial, insurance, real estate, and business service sectors have outpaced those in other sectors; levels of manufacturing employment and wages have both plummeted in recent years.[12]

Furthermore, as in Toronto and Vancouver, Montreal's labor force includes employees from a wide range of ethnic backgrounds. Though 59.9 percent of the metropolitan Montreal's 1991 labor force was French by ethnic origin, no other single ethnic group could lay claim to a position of secondary labor force dominance.[13] The result is an exceedingly fragmented and differentiated income structure crisscrossed by multiple linguistic, ethnic, and gender fissures.

Montreal's increasing diversity has not totally eroded the broad patterns of the past. Though transnational migrants to the city generally find stable employment within a few months, male migrants from elsewhere in North America and Europe generally do so within three weeks of arriving. Men from Sub-Saharan Africa and Asia, for their part, require twenty weeks to secure their initial jobs. Migrants from Sub-Saharan Africa and Asia report the city's lowest income levels (even those with high levels of education who secure employment in higher-status jobs), while non-Canadian North American and European migrants quickly become high-income earners.[14]

Montreal's labor market reveals other distinctive characteristics as well. For example, total employment increased in Montreal between 1971 and 1991 at a much slower pace than in Toronto and other major Canadian employment centers.[15] Montreal simultaneously lost ground to Toronto and some other major Canadian cities in the lucrative finance, insurance, and business service sectors. This pattern persisted into the following decade, despite the fact that Montreal had become home to the highest proportion among all major Canadian cities of residents holding university degrees.[16] By the 2001 Canadian census, the average income of full-time employees in Montreal ($41,763) lagged behind that found in several other Canadian cities (many of which had retained greater opportunities for employment in finance, insurance, and business), such as Ottawa ($53,520), Toronto ($50,516), and Calgary ($49,018).[17]

Moreover, Montreal has remained relatively more dependent on public-sector employment than other major Canadian cities (a pattern that places noncitizen migrants at a disadvantage when seeking jobs).[18] Quebec's laws and union customs greatly inhibited transnational migrant access to construction jobs, a sector dominated by nonnative labor elsewhere in Canada.[19] Primarily, however, Montreal's employment patterns—like those in education—are distinguished from the rest of metropolitan Canada by the issue of language.

François Vaillancourt's analyses of income in relation to language skills during the 1970s and 1980s point in the direction of Montreal's numerous distinctive internal divisions, a sociopolitical segmentation that becomes increasingly visible when viewed through the prism of labor rather than real estate markets.[20] Vaillancourt identified a finely grained income hierarchy in which bilingual anglophone men earned the highest average income of any group in the Province of Quebec.[21] The results of his research, which was conducted for the provincial government's Council on the French Language (Consiel de la Langue Française), were striking for the complex linguistic distinctions that had entered into Quebec's political and scientific discourse by the 1970s.

Vaillancourt's work identified a sharp gender income divide (which proved to be consistently more pronounced than either linguistic or ethnic distinction), and it demonstrated the extent of income inequality throughout the Montreal labor market of the period (with franco-lingual allophone women—e.g., transnational migrant women who speak French—receiving just 38 percent of the income earned by bilingual anglophone men). Research such as that by Vaillancourt, appearing as it did at the heights of debates over language policy and legislation, sounded an alarm to many throughout Quebec that the French language was well on its way to losing pride of place in the province's economic life. Therefore, many concluded, the French language required the protection of special legislative mandates and remedies.[22]

Vaillancourt continued on to argue that the use of language at work was further differentiated by economic sector. Firms engaged in the manufacture and/or sale of goods and services primarily for an internal Quebec market privileged francophones, while the ability to communicate freely in English was more highly rewarded in those firms oriented toward markets external to Quebec.[23]

Vaillancourt thus demonstrated that those who were bilingual in both French and English were paid more than unilingual speakers of either language; that those who were Canadian born earned higher incomes than those who had moved to the province from abroad; and that males' wages were higher than females' wages throughout Quebec (including Montreal). Whereas native-born Canadian women subsequently advanced in both public administration as well as in the business service and financial sectors (especially in high-paying, bilingual professions), their expanding presence in high-earning jobs has done little to alter the place of transnational migrant women at the bottom of Montreal's employment hierarchy.[24]

Recent labor market research in Montreal largely supports McAll's contention that language politics has steadily reconstructed the world of work throughout Montreal since the 1960s. According to McAll, two processes have exercised countervailing influences over the Montreal workplace. On the one hand, the province's language laws enacted by sovereigntist (souverainiste) governments elevated French to the status of the sole official language for public discourse throughout Quebec. On the other, English continued to consolidate its hegemonic position as the language of transnational communication throughout North America and around the world.

These trends place French in a position of near monopolistic dominance throughout the city's extensive public sector. A 2005 census of Montreal municipal employees, for example, revealed that cultural minorities whose mother tongue is neither French nor English fill 5.86 percent of all municipal jobs even though they constitute 29 percent of the city's population, whereas women are only 39.1 percent of the municipal workforce but represent 52.1 percent of the population.[25] French is omnipresent in lower-level management ranks, especially in manufacturing, and it has become the workday language of blue-collar workers as well. English retains pride of place in upper-level jobs in the business, scientific, and financial sectors despite an increasingly informal French-language social environment in the workplace.[26]

In other words, both dominant language groups have come to occupy separate spaces within the Montreal economy and labor market. Transnational migrants, who once favored English, increasingly have gone "to great lengths to learn French (in order to escape from low-paid work with low language content) only to emerge onto a labour market in which employers frequently require competence in English as well as, or rather than, French."[27] Those anglophones who have chosen to remain in Montreal learn French as well. Anglophones and allophones alike reside in multiple linguistic worlds: those of their own families and kin, the neighborhood, the classroom, and the workplace. Interestingly, data on language achievement and anxiety reveal that both groups acquire French with equal agility and asperity.[28] The intricate divisions of Montreal's residential neighborhoods pale in comparison with the linguistic perplexities of the city's and region's labor markets and classrooms.

Concurrent with a linguistic redefinition of the classroom and work-place, Montreal—like older industrial cities throughout North America—has undergone a fundamental restructuring of its metropolitan economy. Many "headquarters" jobs moved to Toronto as major corporations fled the perceived threat of the province's pro-French-language laws. Simultaneously, Montreal's traditional manufacturing base collapsed, entangling trans-national migrants in an unpredictable universe that they often could not understand, let alone control. The disappearance of factory jobs generally drove up the city's unemployment rate, forcing transnational migrants and local blue-collar workers alike into lower-income service-sector employment.[29]

During the last third of the twentieth century, Montreal lost its centrality within the Canadian corporate economy as many major private companies—especially in the financial sector—shifted their headquarters operations to Toronto. The consequent flight of upper-level English-language managers to Ontario was significantly offset by the emergence of a new French corporate management class—as McAll puts it, a sort of "Quebec, Inc."—which rose to the top of major public-sector ("Crown") corporations, as well as provincial government bureaucracies. High-income jobs shifted from anglophones to bilingual anglophones and native francophones as those companies that remained in Montreal increasingly focused on a Quebec provincial—rather than a Canadian national and North American international—hinterland.[30] This labor market transformation has tended to place well-educated and professional transnational migrants at a disadvantage, with the consequence that Montreal's transnational migrants earn less and have a higher rate of unemployment than indigenous Canadians, even though they have higher levels of university education than nonimmigrants.[31]

Manufacturing headquarters eventually followed the financial sector to Toronto, though with a slight delay behind the departures of top financial and business service corporate offices.[32] The city nonetheless retained a robust labor market in specialized manufacturing facilities, with particular strength in aeronautics and aerospace, telecommunications equipment, and biopharmaceuticals.[33] By the mid-1990s, for example, Montreal had become the top-ranking North American metropolitan area in the ratio of the number of high-technology jobs per capita in these fields, as well as the continent's ninth-ranking metropolitan area in the number of high-technology firms with more than 100 employees.[34] Such "new age" industries, together with lower-paying jobs in the service sector, have provided employment opportunities for the city's substantial transnational migrant population.

Furthermore, Montreal has been able to leverage its unique position as a French-language metropolis in the Americas to attract a significant number of headquarters for international organizations. The city has begun to use its distinctive linguistic environment to create institutions for accommodating diversity, to increase diversity capital. By 2000, it had attracted the central offices of nearly fifty international agencies, making it second to Washington as a preeminent administrative center for such institutions.[35] Yet as important as these offices may be for the city's general prestige, such employers do not add significantly to the job opportunities for most transnational migrants.

The emergence of a francophone labor market combined with strong public-sector economic involvement sustained the city's vibrant arts and cultural industries, despite the departure of some major film, television, and radio studios.[36] The consequent arts scene has reinforced Montreal's attraction as a tourist destination (constituting another sector offering employment opportunities to transnational migrants, though, once again, in predominantly low-wage jobs).[37]

Meanwhile, Montreal has been able to continue to capitalize on an extensive educational, life sciences, and health sector. Specialized research facilities demand fluency in the dominant international language of scientific communication: English. Major institutions in these fields offer a wide range of employment options, with more menial support services often being filled by immigrants.[38]

Taken in the aggregate, the last decades of the twentieth century—a time of significant political upheaval in Montreal as well a moment of arrival for tens of thousands of migrants from abroad who increasingly traveled to Canada and Quebec from Africa, Asia, and the Caribbean—witnessed a profound restructuring of the city's metropolitan economy. This dislocation reflected both a broader continental trend toward postindustrial employment together with specific features resulting from the city's complex language politics. As summarized in table 3.1, the metropolitan Montreal labor force expanded by nearly a half million, with the number of jobs growing in managerial and professional ranks, remaining largely stable in the sales and service sectors, and decreasing in more traditional industrial supervisory and blue-collar categories.

Such an explosive combination of cross-cutting trends fragmented the city's labor market at the same moment when the historic city's residential patterns were disintegrating. Language mattered more and more as it determined economic success (though not in quite the straightforward manner reflected in the political debates of the era). Economic life became increasingly segmented, with factors such as language, gender, profession, economic sector, and educational attainment working to subdivide and segment employment opportunities into ever smaller niches.

*Table 3.1. Occupational Structure, Montreal Census Metropolitan Area (1981 boundaries), 1971, 1981, and 1991 (percent)*

| Occupational Category | 1971 | 1981 | 1991 |
|---|---|---|---|
| Managers | 3.5 | 5.9 | 7.5 |
| Professionals | 12.7 | 12.6 | 15.1 |
| Supervisors | 9.9 | 8.8 | 8.4 |
| Upper-level white-collar/technical | 18.9 | 20.7 | 21.2 |
| Lower-level white-collar/sales/ service workers | 29.2 | 28.5 | 28.7 |
| Skilled blue-collar production workers | 12.4 | 10.9 | 9.7 |
| Semiskilled and unskilled blue-collar production workers | 13.4 | 12.6 | 9.4 |
| Total number of people | 953,375 | 1,348,055 | 1,509,140 |

*Source:* Statistics Canada, Censuses of Population, special compilations for D. Rose and P. Villeneuve, as presented in Annick Germain and Damaris Rose, *Montréal: The Quest for a Metropolis* (New York: John Wiley & Sons, 2000), 141.

A dissonance of cultural difference has come to dominate ever-more-finely defined gradations of school life as well.[39] In one particularly dramatic incident that played itself out to considerable media attention, a twelve-year-old Sikh boy from the Punjab, Gurbaj Singh Multani, ignited controversy in late 2001 by claiming a right to carry his *kirpan* (a ceremonial curved dagger that Sikh males must wear at all times) to school. Multani and his ultraorthodox Sikh parents turned to the judicial system, where they won an initial court case, inciting deep resentments among the parents of his classmates in the process.[40] The Quebec government and the local school board appealed the lower court's decision, with the Quebec Court of Appeal ruling against Gurbai and his family in March 2004.[41] Gurbai's family immediately sought remedy from Canada's Supreme Court.[42] The country's highest court placed the case on its docket seven months later, with hearings and decisions expected some time in 2005.[43]

The Singh incident recalled an earlier case of cross-cultural conflict in Montreal classrooms over appropriate attire. Émilie Ouimet, a teenage student at a local public school, successfully petitioned provincial authorities in 1995 for permission to wear the *hijab,* the Muslim female head covering, to class. The Ouimet decision resolved the question of Muslim attire in public schools, although the issue resurfaced in a private school beyond direct provincial authority in September 2003.[44] Commenting on the 2003 case, Ginette L'Heureux of the Quebec Human Rights Commission and other human rights advocates observed that *hijabs* were no longer controversial in "mainstream" Quebec society.[45]

Such cases—and the attention surrounding them—reveal the extent to which Montreal has been re-created as a multiethnic cultural center in recent years even as Bill 101 has established the legal primacy of French throughout Quebec. The arrival of new migrants from varied cultural milieus around the world has added yet a new dimension to already complex work and school environments. Montreal's intricate "daily 'dance of language'" at work and at school has become accompanied by an equally complex "dance of culture." The city's capacity to adapt to diversity—its diversity capital—has increased.

## Washington: A Southern Government Town No More

Transnational migrants similarly began to arrive in Washington at a time of profound economic change. In the case of Washington, the link between growing in-migration and economic development is perhaps clearer than in Montreal. Briefly put, the Washington metropolitan area has experienced more than three decades of sustained, large-scale economic growth, transforming the city and region from a one-industry (government) company town into an increasingly diversified metropolitan-scale regional economic powerhouse.

Washington is, in many ways, a quintessential postindustrial city that generates wealth through the exchange of information and services rather than physical goods.[46] Unlike Montreal, Washington did not have an extensive industrial base that needed to give way to a postindustrial socioeconomic reality. Manufacturing never played a significant role in the Washington economy, which is one reason why the region had not been especially attractive to transnational migrants before the latter decades of the twentieth century. Those transnational migrant communities that did exist in Washington were employed first and foremost in small-scale commerce.[47] Washington's blue-collar working class, to the extent it existed at all, filled the ranks of the region's construction industry, kept stores and private offices clean and functioning, and provided valuable support services for the government.

As the data in table 3.2 indicate, Washington's recent economic transition has differed from what has transpired in more traditional industrial cities such as Montreal. The main shift in the Washington economy has been an evolution away from a hegemonic public service sector to a diverse regional economy in which private firms have come to play an ever larger role (though many private-sector jobs are tied in some manner to the presence of the federal government).[48] Another profound development has been the relative diminution of employment opportunities within the District of Columbia, accompanied by the explosive growth of job opportunities in the suburbs. Between 1971 and 2001, total employment in the city grew from 567,000 to 651,000 jobs. Meanwhile, during the same period, according to the Bureau of Labor Statistics, the total number of jobs in the Washington metropolitan area exploded from 1.2 to 2.8 million.[49]

In other words, employment opportunities were more or less equally divided between city and suburb on the eve of significant transnational migration to the Washington region in the early 1970s. Within just thirty years, migrants were entering a regional economy in which only one job in four was within the borders of the District of Columbia. Virginia's Tyson's Corner and Maryland's Bethesda, along with Virginia's Dulles Corridor and Maryland's parallel Interstate 270 Corridor, literally defined what would be-

*Table 3.2. Structure of Employment by Sector in Washington and Its Metropolitan Area, 1991 and 2001 (percent)*

| | 1991 | | 2001 | |
| Sector | Washington | Metropolitan Area | Washington | Metropolitan Area |
|---|---|---|---|---|
| Government | 41.5 | 27.6 | 34.2 | 22.1 |
| Services | 37.6 | 33.8 | 46.7 | 40.6 |
| Wholesale and retail trade | 8.5 | 19.3 | 7.9 | 17.7 |
| Finance, insurance, and real estate | 5.1 | 5.8 | 5.1 | 5.4 |
| Construction | 1.6 | 4.8 | 1.7 | 5.7 |
| Transportation and public utilities | 3.5 | 4.5 | 2.7 | 4.9 |
| Manufacturing | 2.2 | 4.2 | 1.7 | 3.6 |
| Total number of people | 677,300 | 2,291,800 | 650,900 | 2,794,300 |

*Source:* Bureau of Labor Statistics, U.S. Department of Labor, ftp://ftb/bls/gov/pub/special.requests/ philadelphia/fax_9554&6/txt.

come widely known as the "Edge City" (this term having been coined by the *Washington Post* writer Joel Garreau in reaction to all that he was seeing around him as these sprawling exurbs became his "beat").[50] By 2003, more than half the Washington metropolitan region's workforce was commuting across jurisdictional lines to their jobs.[51]

Transnational migrants followed jobs, so distinct migrant communities have come to occupy employment niches tied to specific subregions within the metropolitan area.[52] For instance, South Asian medical researchers and computer specialists tend to work in the high-technology employment centers around Dulles Airport, and biotechnology employers huddle around the National Institutes of Health campus across the Potomac River near Bethesda. Salvadoran construction workers are to be found in every corner of the Washington metropolitan region, as are African cab drivers.

With different migrant groups drawn to specific labor submarkets throughout the region, Washington has become a distinctive migrant-receiving region in that no single migrant group predominates. As noted in the previous chapter, El Salvador has been the leading country of origin for Washington area transnational migrants, with 10.5 percent of all arriving legal immigrants in the Washington region; it is followed by Vietnam, with 7.4 percent; and India, with 5.5 percent; no other single country contributes more than 5 percent.[53]

Audrey Singer and her colleagues report in the Brookings Institution study of immigration to greater Washington (cited above in chapter 2) that

> between 1990 and 1998, nearly one-quarter of a million immigrants from 193 countries and territories chose the Washington metropolitan area as their intended residence. The majority of the newcomers are in their prime working years, and thus are an important supply of new labor. Fully 75 percent are 40 or younger, and the mean age of the immigrant population is 29. Twenty-five percent of the recent immigrants are under 18 years old. Of these recent arrivals to the metropolitan area, 53 percent are female and 47 percent are male.[54]

Moreover, transnational migrants, like the jobs which they pursue, have spread themselves out throughout the region. Singer and her colleagues find that four of the region's top ten immigrant destination postal zip codes are located in Virginia, four in Maryland, and two in the District of Columbia.[55] Thus, Washington has become one of North America's first leading twenty-first-century transnational migrant centers, a region where residents encounter tremendous diversity every day. The workplace and schoolyard are where people of different backgrounds come together most often.

As in Montreal, income hierarchies for native and transnational residents are highly fragmented and uncommonly complex in the Washington metropolitan region. Research focusing on various migrant groups in the region is scant. This striking absence of reliable data appears in large measure to be an artifact of the region's myriad jurisdictions, which do not combine neatly into census reports. These statistical lacunae are amplified by an apparent underappreciation by all too many local leaders of the intense population changes sweeping the region. As the Council of Latino Agencies laments in its 2002 report *The State of Latinos in the District of Columbia:*

> Limitations in data collection, a lack of consistency in population identification, and a tendency to conflate and overlook variables of ethnicity and race limit in-depth analysis of many key socio-demographic characteristics, particularly trend analyses over time. These data limitations severely restrict research and analysis on Latinos [*author's note:* and other transnational migrant communities] and hinder meaningful interpretation of complex socioeconomic patterns.[56]

Analyses by Robert Manning based on data collected for the late 1980s go the farthest in capturing Washington's increasingly intricate income distribution. Manning found that

> overall, African American families are more than three times as likely as White families to live in poverty and over three times less likely to earn over $75,000 in 1989; Hispanics are slightly better off, followed by Asians. In terms of median household income, whites ($46,000) earn over $21,000 more than blacks ($25,000) in the District with Hispanics ($26,000) and Asians ($30,000) relatively close behind. Although the suburbs evidence substantially lower rates of poverty and much higher incomes, the most striking pattern is the consistently wide disparity in household income across all race and ethnic groups; the exception is the doubling of Asian income (over $70,000), which exceeds that of whites in the suburbs. On average, median household income in the suburbs increases by about $14,000 for Hispanics, $21,000 for Native Americans and blacks, over $20,000 for whites, and a remarkable $40,000 for Asians.[57]

Several more recent, yet incomplete, sources confirm the overall economic hierarchy identified by Manning more than a decade ago. Transnational migrants are playing an ever more visible role in the creation of new firms throughout the Washington metropolitan region. Migrants thus generate capital and jobs for the regional economy at rates faster than African-Americans, though more slowly than firms owned by native-born whites.

In this regard, the Greater Washington Ibero-American Chamber of Commerce reported that 32,000 Hispanic businesses were operating throughout the Washington region in 2002, as opposed to just 500 thirty years before.[58] Migrant businesses range from small "mom-and-pop" restaurants such as the Husacaren Restaurant on Mount Vernon Avenue in Alexandria, Virginia,[59] to large media companies such as Arlington, Virginia's, WZDC Television, Channel 64, a Spanish-language cable TV company employing 160 employees in several states,[60] and Thie Le's String Bean Software firm in Gaithersburg, Maryland, which supplies its products to Microsoft Corporation.[61] Analysts for the Council of Latino Agencies found in 2002 that 377 Latino firms operating in Washington employed more than 4,300 workers and generated more than $600 million in goods and services. The Council of Latino Agencies suggested further that many Central and South American migrants to the city have managed to occupy more secure niches in its over-all economy.[62]

The council's writers continued on, adding that unemployment among the city's Latinos had fallen throughout the 1990s from 6.8 to 5.6 percent, in comparison with a white unemployment rate of just 2.1 percent and a black rate of 8.2 percent in 2000. The authors nonetheless warned that the September 11, 2001, terrorist attack on the Pentagon disproportionately af-fected low-wage hospitality workers, an employment sector dominated by Latinos. Consequently, the Latino unemployment rate soured to more than 10 percent following 9/11, even as unemployment held steady for the city's whites and blacks.[63]

Later reports have indicated that any effects of the 9/11 events proved to be short lived. An analysis of census data conducted by the Center for Im-migration Studies, for example, revealed that immigrants accounted for one in five jobholders in the Washington metropolitan area by the end of 2003.[64] Employment rates among the region's foreign born rose even as the region's overall general unemployment increased simultaneously. Moreover, the Center for Immigration Studies confirmed that transnational migrants held jobs throughout the regional employment hierarchy, "from office cleaners to computer programmers."[65]

The economic success of the Washington area's Latin American com-munities adds fresh dynamism to the regional economy. More particu-larly, 3,000 Salvadoran-owned small businesses (e.g., restaurants, retail stores, and construction companies) enhance life in many central-city and inner-suburban commercial areas long abandoned by more established Washington entrepreneurs. Such areas include Washington's Adams Morgan; Maryland's Langley Park and Wheaton; and Virginia's Bailey's Crossroads, Woodbridge, Manassas, Fredericksburg, and Arlandria.[66]

Hints of the growing scale of Hispanic economic activity may be found in data on remittances sent to families back home. According to the Inter-American Development Bank's estimations of remittances in 2004, Latin American immigrants in Maryland led the nation in funds sent to their country of origin ($2,894).[67] Latin American immigrants living in Virginia ranked fifth among transnational migrant communities nationwide ($2,617), and their neighbors in Washington came in sixth ($2,629). Over-all, the bank estimated that Latin Americans living in the District of Co-lumbia, Maryland, and Virginia would send $1.2 billion outside the United States in 2004.

Despite the inadequacy of employment data, such alternative indicators suggest that, as in Montreal, newly arriving transnational migrants to Washington enter a highly complex and robust labor market in which they inevitably encounter their city's and metropolitan region's growing diversity. New Washingtonians learn to navigate an intricate world of race relations every time they show up for work. As Manning's work reveals, the ever-nuanced world of race in Washington is about economics as well as about etiquette. Whites, on average, earn nearly twice as much as blacks. African-Americans are three times more likely to live in poverty, and blacks remain three times less likely to have high incomes as whites.

Washington's transnational migrants fall in between these "founding" groups. The new middle ground among Washington area blacks, whites, and transnational immigrants is becoming ever more visible in the region's schools. Schools, therefore, join the workplace and the neighborhood as venues within which people of diverse cultures come together. They also provide environments within which one might expect new diversity capital to emerge.

Data concerning the racial and ethnic composition of Washington area public schools accentuate the region's diversity, because many affluent parents in some jurisdictions opt out of government schools for private educational alternatives. Even so, one indisputably significant fact is that classrooms throughout the region's various public school systems are entering the twenty-first century with a variety of students that would have been unimaginable a generation before.

As the data in table 3.3 reveal, white students constitute a majority of students in only one of the region's six major public school systems (Fairfax County, Virginia), with African-Americans representing a majority among public school enrollees in two districts (District of Columbia and Prince George's County, Maryland). Asian, Pacific Islander, and Hispanic students (often the children of the transnational migrants whose parents are filling the new private service-sector jobs described above) combine to form the largest group in one district (Arlington County, Virginia). However, this ethnically and racially mixed statistical category now constitutes the second largest group in all the remaining public school systems in the region (District of Columbia; the city of Alexandria and Fairfax County in Virginia; and Montgomery and Prince George's Counties in Maryland).

Age data by racial and ethnic group suggest that such school enrollment patterns are unlikely to change in the future. For example, in 2000, 91 percent of all white residents of the District of Columbia were over the age of

*Table 3.3. Racial and Ethnic Group Composition of Student Population in Washington Metropolitan Area Public School Systems, 2003–4 (percent)*

| Racial and Ethnic Group | District of Columbia, Oct. 7, 2003 | Alexandria, Va., March 15, 2004 | Arlington County, Va., March 15, 2004 | Fairfax County, Va., March 15, 2005 | Montgomery County, Md., Sept. 30, 2004 | Prince George's County, Md., Sept. 30, 2004 |
|---|---|---|---|---|---|---|
| White | 4.86 | 22.95 | 41.7 | 52.76 | 44.6 | 8.1 |
| African-American | 83.60 | 42.90 | 14.4 | 10.68 | 22.1 | 77.6 |
| Asian or Pacific Islander | 1.73 | 6.69 | 10.2 | 16.8 | 14.3 | 3.1 |
| Hispanic | 9.75 | 26.98 | 32.8 | 15.099 | 18.79 | 10.8 |
| Native American or Alaska Native | 0.05 | 0.27 | 0.099 | 0.37 | 0.00 | 0.6 |
| Multiracial | 0.00 | 0.00 | 0.00 | 0.0 | 0.00 | 0.00 |
| Undesignated | 0.00 | 0.02 | 0.14 | 4.29 | 0.00 | 0.00 |
| Total number of pupils | 65,099 | 10,902 | 19,158 | 164,235 | 140,000 | 137,285 |

*Sources:* District of Columbia Public Schools, "Summary of Membership Report," October 7, 2003, http://www.k12.dc.us/dcps/ frontpagepdfs/membership Oct703_race.pdf. Virginia Department of Education, "Student Membership Report, September 30, 2003," http://www.pen.k12.va.us/VDOE/dbpubs/Fall_Membership/2003/readme.html. Montgomery County Public Schools, "Annual Report, Fall 2003," http://www.mcps.k12.md.us/departments/publishingservices/PDF/2003annualreport.pdf. Prince George's County Public Schools, "2004 Enrollment Report," http://msp.msde.state.md.us/enroll.asp?K=16AAAA. The author acknowledges the assistance of Janet Mikhlin and Oksana Klymovych, research interns at the Kennan Institute, in the preparation of this table.

18, as compared with 75 percent of African-American and Latino Washingtonians.[68] Moreover, the class of 2003 was the last majority white graduating class from Montgomery County schools, a system that had been 94 percent white just thirty-five years previously.[69]

These data reveal that, as at the workplace, a vast majority of Washingtonians head off to school to spend a good portion of their day with others who are not like themselves. Fewer and fewer individual schools in several of the region's major public school systems can be defined by race, as was the case in the past. More and more school children share classrooms and playgrounds with classmates from quite diverse racial, ethnic, and linguistic backgrounds.

Both the systemwide data reported here, and the individual school statistics on which these data rest, show that race remains a defining element in the District of Columbia and in Prince George's County and, to a lesser extent, in Alexandria. Furthermore, white parents often chose to send their children to private educational institutions rather than to African-American–dominated public schools. As in Montreal, the "daily 'dance of race'" continues even as new groups arrive.

Employment, income, and public school enrollment data demonstrate that the poisonous legacies of slavery and racism still cast a dark shadow on Washington workplaces and classrooms. African-Americans earn less than other groups of Washingtonians. White parents have largely abandoned public school districts such as those of the District of Columbia and Prince George's County in which African-American students predominate. That said, many local whites encounter black colleagues at work who are better educated, earn higher salaries, and have higher ranking positions than they.

The workplace and schoolyard have become primary meeting grounds for all groups in Washington to meet with others who are somehow different. Migrants from abroad not only enter an unfamiliar world; they also change that world in return. A combination of sensational economic expansion, profound economic restructuring, and dramatic metropolitan sprawl has converted small-town Washington into a globalized urban region. Even more than residential configurations, evolving employment and school patterns are transforming race relations throughout the metropolitan region from a bipolar contest of blacks and whites into a multidimensional middle ground of many hues. Workplaces and schools are some of the spaces within the metropolis where diversity capital is formed.

## Kyiv: Informal Markets and After-Hours Classes

Sorting out group employment patterns in Kyiv is a more difficult task than in Montreal or in Washington, because official statistics either do not exist or fail to inspire confidence. Far too great a proportion of the local economy falls outside official record keeping to permit more than speculation about how many people work at what sorts of jobs, with whom, and at what location. The scale of Ukraine's "informal" economy is unknown, with estimates running as high as 60 percent of total economic production.

Foreign migrants arriving with only the most tenuous connection to local officialdom naturally gravitate to those professions that serve the country's enormous "gray" economy. Kyiv's migrants have been sorting themselves into distinct economic niches by ethnicity and national origin. For example, 92.9 percent of employed Pakistani respondents to a 2001 Kennan Institute / Kennan Kyiv Migrant Survey were engaged in trade or were self-identified entrepreneurs, as were 85.4 percent of working Afghan respondents, 72.7 percent of the economically active Vietnamese respondents, and 71.4 percent of employed and self-employed respondents from the Middle East. Each national group struggled to carve out a separate and identifiable trading sector for itself.

By contrast, 66.6 percent of African respondents to the survey were employed as blue-collar workers.[70] This marked difference in African employment (and also residential) patterns may well be a consequence of race. Though racial difference undoubtedly becomes a factor separating Africans from other transnational migrants, more research is required before it will be possible to fully assess the role of visible difference in shaping the experience of Africans in Kyiv.

Differentiation among migrant groups is more pronounced in the self-reporting of monthly incomes by respondents from various national backgrounds. The acknowledged earnings of African respondents in the Kennan Kyiv Migrant Survey, for example, are just 22 percent of the higher monthly incomes reported by Chinese respondents. Pakistani, Middle Eastern, and Vietnamese migrants reported earnings more closely approximating those of the Chinese survey participants (77.2, 86.7, and 92.2 percent, respectively).[71] Though the higher reported earnings of Middle Eastern and Vietnamese migrants may be a result of their longer residency in Ukraine (and the Soviet Union before 1991), Chinese survey participants are relatively recent arrivals in Ukraine. Some migrants were traders and entrepreneurs at home; most, by and large, were not. The self-proclaimed traders and entrepreneurs in the survey sample were most often former students, workers, military personnel, and office workers. Hence, higher earnings may be related to higher professional and educational achievement in a migrant's country of origin.[72]

Migrant traders at local markets, and Kyiv's migrant entrepreneurs more generally, appear to have taken up their new professions as a consequence of necessity during migration. Many migrants draw on the general skills and connections that they brought with them from abroad to enhance their efforts in small-scale trade in Ukraine. Migrants interact with the native Kyiv population through trade (many customers coming from outside their own migrant community), and through their interaction with officials. Relations between migrants and local officialdom often appear to reinforce negative stereotypes on both sides. For their part, local Kyiv police frequently consider migrant traders as not really working but as merely "hanging out in the markets."[73]

Although a majority of Kyiv's migrants report being employed in an informal trading economy, some have begun to establish small factories. Phan Viet Hung, for example, has taken over the floor of a semiabandoned factory on the Ukrainian capital's northwest fringe to set up a printing and photography business that relies on contracts with Fuji Film.[74] Phan first came to the Soviet Union during the 1960s to study in Moscow and has stayed on ever since. Married to a Ukrainian, he ended up in Kyiv following the 1991 collapse of the Soviet Union. By 2000, he was running his small printing operation with about two dozen Vietnamese, Russian, and Ukrainian employees. Once again, it was in the workplace rather than at home that migrant and native came together.

Kyiv's classrooms have emerged as an even more important space than the workplace for cross-cultural contact. The absence of special policies governing foreign students in Ukraine's public schools has proven to be one major factor contributing to the integration of the city's various migrant and ethnic groups into city life more generally. Underfunded—and concerned with issues of state building—the national Ministry of Education rarely acknowledges the existence of groups of transnational migrant children in the nation's schools. Consequently, the children of Kyiv's Afghan, Kurdish, African, and Vietnamese migrants attend schools that offer no special programs encouraging acculturation, provide no special language training in either Russian or Ukrainian, and organize few remedial courses on other subjects. Interestingly, teachers, school administrators, migrant parents, and the children themselves report that the general academic performance of the students whose parents have been born abroad is indistinguishable from that of children whose parents were born in Ukraine.

Such primitive "mainstreaming" of migrant children is a consequence of educational policies and constitutional provisions that seek to ensure free access to a general education for every child in Ukraine.[75] The right to education extends to all children native to Ukraine, to all children of legal immigrants, and those children whose parents have obtained official refugee status. Consequently, the children of parents who cannot provide documentation demonstrating legal residence in Kyiv generally may be excluded from school.

Although the enforcement of such provisions varies considerably from one school to the next, the end result is that significant numbers of children of transnational migrants do not attend local schools. For example, nearly one-quarter of the households participating in the Kennan Kyiv Migrant Survey were unable to enroll their children in school. These children make their way into the informal economy, tending to remain unassimilated into Ukrainian life.

Generally, however, the classroom becomes a major arena for assimilation into city life and culture for the vast majority of migrant children who attend Kyiv's public neighborhood schools. As was noted above, migrant students perform at levels of achievement similar to those of children native to Ukraine. Three-quarters of the children from families included in the Kennan Kyiv Migrant Survey enrolled in classes appropriate for their age. School attendance rates among migrant children enrolled in school are high (perhaps because nearly all those surveyed are eligible for the free breakfast, lunch, and textbook programs available to all low-income children in the Kyiv public school system).

The extent to which Kyiv's schools advance the assimilation of transnational migrant children into local life and form diversity capital may be seen in the language use patterns of those same children. About two-thirds of non-Afghan migrant children surveyed by the Kennan project were enrolled in Ukrainian language schools, while a similar percentage of Afghan children enrolled in Russian language school. The vast majority of children who arrived in Kyiv before 1995 attend Ukrainian-language schools, though this trend was reversed in favor of Russian-language schools during the late 1990s.

Overall, 97 percent of migrant children participating in the Kennan survey claimed proficiency in Russian, 91 percent in the national language of their parents, and 68.7 percent in Ukrainian. Though 86 percent of the children questioned during the Kennan Kyiv Migrant Survey speak the native language of their parents at home, 31 percent also speak Russian and 8 percent speak Ukrainian when they are in private with their relatives. However, once away from their parents, 70 percent of children speak Russian with their friends, and another 20 percent converse in Ukrainian.

Ukrainian-language usage among transnational migrant children, though lagging behind that of Russian and of parents' native languages, is higher than the levels reported in surveys of nonethnic Ukrainians living in Ukraine in the early 1990s. During Ukraine's postindependence years, various studies reported that non-Ukrainian native speakers (ethnic Russians, Jews, Tatars, Hungarians, et al.) rarely used Ukrainian outside environments in which it was officially mandated.[76] As in Montreal, transnational migrant children in Kyiv are establishing their own multilingual environment in which all the participants use three languages interchangeably—that of their parents, Russian, and Ukrainian.

Schools similarly advance the assimilation of parents, enhancing Kyiv's diversity capital. Given the general impoverishment of all Ukrainian schools, parents are expected to assist in the maintenance and operation of their children's schools in some meaningful way—either with under-the-table cash payments or with free labor and in-kind contributions. Among those parents surveyed, 96 percent report having ongoing contact with their children's schools, often through community service and parent meetings.

Administrators at several individual schools, in turn, have made their facilities available to local migrant communities for special evening and weekend language and culture courses. A total of 100 percent of Vietnamese, Chinese, Middle Eastern, and South Asian parents responding to the Kennan Kyiv Migrant Survey reported that their children attend after-hours language and cultural training courses in their native language. Afghan and African respondents reported slightly lower, though impressive, participation rates in community-based supplemental education opportunities (80 and 70 percent, respectively).

Kyiv's schools form a critical middle ground into which the children of immigrants and their parents come into direct contact with Ukrainian society (and with other migrants as well). Even more than the world at work, the school has become as the preeminent venue for interaction among Kyiv's burgeoning migrant communities. Many Kyiv schools have become the place in which the city's international future has already become its present.

## Work, School, and Diversity Capital

In Kyiv, as in Montreal and Washington, the manner in which transnational migrants come together with the indigenous populations—French and English, black and white, Ukrainian and Russian—is determined by where and how they work and go to school. More than at home and in their neighborhoods, Montrealers, Washingtonians, and Kyivans find themselves interacting with others not like themselves as they make their way through the schoolday and workday. The workplace, the school, and the playground have become leading meeting grounds in which adult migrants and their children establish a place for themselves between more traditional groups and their entrenched divisions.

Montreal, Washington, and Kyiv are quite different metropolitan communities today from what they were a generation or two ago. They have become the focal points of transnational migration systems that bring thousands of new residents from across international boundaries into local neighborhoods, workplaces, and classrooms. All three cities share more with traditional entrepôts such as New York and Amsterdam at the dawn of the twenty-first century than was the case at the beginning of the twentieth. They represent cities where long-standing hostility and firm boundaries between linguistic and racial groups are becoming increasingly blurred by the arrival of new residents, coworkers, and schoolmates who do not fit into earlier conceptual maps of the metropolis. These processes suggest some of the ways in which new diversity capital may be created and accumulated.

Diversity capital, however, is not merely the coming together of people of diverse backgrounds. The notion implies an expansion of capacity to achieve a characteristic that Richard Stren and Mario Polese have identified as "urban social sustainability." Urban social sustainability, in turn, requires the emergence and maintenance of "policies and institutions that have the overall effect of integrating diverse groups and cultural practices in a just and equitable fashion."[77] Hence, the creation of diversity capital—or a societal capacity to absorb diversity—requires political action.

An exploration of how cities create new capacities for accommodating diversity must examine how the presence of growing transnational communities alters the urban political landscape, because it is ultimately through the political process that policies and institutions come into existence and assume meaning. The next three chapters, therefore, seek to explicate how urban politics changed in Montreal, Washington, and Kyiv during the last third of the twentieth century.

The starting point for a discussion of political change in Montreal and Washington must be the political unrest and communal turbulence of the late 1960s, when both cities endured painful outbursts of civic violence that have echoed throughout the subsequent years. The repression of the Soviet regime muted open unrest in Kyiv at that time. Since the mid-1980s, however, civic demonstrations have been far more frequent in Kyiv than in either Montreal or Washington (leaving aside demonstrations concerning national issues in Washington). Public protests against the Soviet government's handling of the Chernobyl' nuclear accident marked the beginning of a period of large-scale street protests and political demonstrations. More recently, nothing in either North American city approaches the drama, consequence, and scale of Kyiv's revolutionary response to election fraud in the November 2004 Ukrainian presidential election now known as the "Orange Revolution."

As the following chapters suggest, common patterns begin to emerge in which all three urban communities represent different stages in a similar cycle of urban change whereby large numbers of transnational migrants transform previously "divided" urban cultural, social, economic, and political landscapes. This pattern of diversity capital formation is most mature in Montreal, and it may be said to have already profoundly altered how that city functions. Washington represents an instance in which demographic transformation remains largely unnoticed, only recently entering into general public awareness. The processes examined in this volume are in their infancy in Kyiv. Taken together, the recent histories of Montreal, Washington, and Kyiv suggest some of the ways in which cities and metropolitan regions elsewhere might begin to increase their capacity to absorb diverse transnational populations and enhance "urban social sustainability."

# Part III

# Political Transformations

# Chapter 4

# From Saint Jean-Baptiste to Saint Patrick: Montreal's Twisting Path to Intercultural Diversity

> When it is minus 20, or even just minus 8, but with slush and gray skies, one asks oneself, "Why am I in Montreal and not Rio or Zanzibar?" "It's because it is an extraordinary town, c'est une ville formidable," we respond in unison. "Have you forgotten?"
>
> —Marie-Claude Lortie, Jean-Christophe, Laurence, and Rafaele Germaine, "Montréal qu'on aime," *La Presse*, February 1, 2003

He sat impassively by himself amid smoke and flying debris in the viewing stand's front row along the street running past Lafontaine Park.[1] He stared out onto the street with a steely gleam that most of his compatriots had never seen before—but would encounter over and over again in the years ahead.[2] All the other dignitaries were gone, but he refused to leave. He remained firmly seated with a handful of police watching the Saint Jean-Baptiste Day parade pass by on a cool June evening in 1968.[3]

"It was a night of frenzy rarely seen in Quebec," commentators would later note.[4] Quickly dubbed "Nightstick Monday" (*lundi de la matraque*), during the night 290 demonstrators would be arrested, and 130 among those present injured. The demonstration's primary organizer—Quebec separatism's most fiery orator, Pierre Bourgault—had managed to unleash a whirlwind that would secure his archrival's reputation as a hero to Canadian federalists.[5] The solitary man in the front row merely sat peacefully and stared impassively long into the Montreal night. The next day that lone spectator, Pierre Elliott Trudeau, would receive what would prove to be his largest mandate as his country's prime minister.[6]

## Saint Jean, the Shepherd Boy

During the 1920s, conservative Catholic Quebec nationalists seized upon the feast day of the French-Canadians' patron saint, John the Baptist, as a pretense to demonstrate their popular authority throughout the province.[7] An annual parade was initiated modestly in 1922, reflecting "a sort of sentimental nationalism with dull speeches dutifully recalling the 1837 Rebellion in Lower Canada."[8] Festivities included floats sponsored by various Roman Catholic parishes and associations, topped off by a central representation of the Baptist portrayed as an appropriately pastoral curly haired shepherd boy.

Twentieth-century Quebec nationalists used the parade to appropriate a long-standing traditional back-country celebration dating from the arrival of the first *habitants* in rural Quebec (and even before, in Normandy). Philippe-Joseph Aubert de Gaspé, in what may have been the first significant Canadian novel in any language (*Les Anciens Canadiens,* first published in 1863), wrote that "rural Canadians had preserved a very impressive ritual handed down from their Normandy forefathers. This was the bonfire set alight at dusk on St. Jean-Baptiste eve, in front of the main door of the church. . . . Crackling flames shot up as the crowd shouted for joy and fired off their guns."[9]

By the mid-twentieth century, the juxtaposition of Saint Jean-Baptiste Day, June 24, with Dominion (later Canada) Day on July 1 drew an ever sharper contrast between Quebec's "two solitudes." French families could luxuriate in various picnics and outdoor festivities on one of the longest days of the year; and they could do so without having to pay obeisance to the anniversary of Canadian Confederation a week later. More significantly, the curly haired shepherd boy represented the mythical agrarian values of the increasingly corporatist regime of Maurice Duplessis.

Maurice Le Noblet Duplessis and his strongly conservative Union Nationale Party tightly controlled Quebec for more than two decades. In alliance with especially reactionary elements within the hierarchy of the Roman Catholic Church, Duplessis's Union Nationale thugs ruthlessly enforced the power and privilege of the province's ruling coalition. Meanwhile, "Le Chef" wrapped himself in an agrarian patriotic vision for Quebec and, by the end of the 1940s, quite literally in a new Quebec flag that his government had adopted in 1948.[10]

Duplessis's primal anti–trade union fervor, combined with the anticommunist populism of his Union Nationale Party, pushed Quebec's contending social forces along a collision course that ended in one of the most bitter strikes in Canadian history at the Thetford and Asbestos mines in Quebec's Eastern Townships.[11] The 1949 asbestos strikes against Canadian Johns-Manville, Asbestos Corporation, Johnson's Company, and Flinkopt Mines became a defining moment in twentieth-century Quebec history—a point in time at which the major forces and figures that would reshape the province over the next half-century aligned together to fight against the reactionary realities of Duplessis's Quebec. Ardent support for the striking miners launched the public careers of the passionate federalists Jean Marchand, Gérard Pelletier, and Pierre Elliott Trudeau. Future longtime Montreal mayor Jean Drapeau established his credentials as a reformer by defending strikers in hostile courtrooms. These events radicalized many young Quebecers who subsequently would embrace more left-oriented forms of prosovereignty sentiment in the years to come, such as René Lévesque (a rising journalist at the time). Thetford proved to be a political nova, spreading its bright debris across the Quebec political landscape for decades to come.

The following year, Pelletier and Trudeau founded the influential journal *Cité Libre,* which would become a leading voice for liberalism, Canadian unity, and antinationalism.[12] Trudeau gained further notoriety with his authoritative study of the strike, *La grève de l'amiante: Une étape de la révolution industrielle au Québec,* which appeared in 1956.[13] The ferment harnessed by *Cité Libre* helped to set the stage for the breakthrough victory of Jean Lesage's Liberal Party following Duplessis death on September 11, 1959. Lesage's win the next year launched a period of intense reform and secularization, which has become known as the "Quiet Revolution."[14]

Lesage's Liberals left no aspect of Quebec life untouched as they wrestled control over education, health, and social welfare away from clerical domination. In so doing, they redefined the very meaning of *québécité.* In a 2003 exploration of values throughout North America, Canadian pollster Michael Adams captured the transformation of Quebec by observing that

> from a society with staunch Catholic roots (implying not only deference to religious and patriarchal authority, from your parish priest right up to the Pope, but hierarchy, belief in the traditional family, duty, and propriety—quite a potent cocktail of traditional values), Quebec has evolved into the most postmodern region on this continent, a status that is a least partially explained by the fact that only 29 percent of Quebecers now believe in the Devil and fewer still, 26 percent, believe in Hell, the lowest

proportions, by far, of any region on the continent—the 3 percent differential apparently made up of those who think the Devil is with us, not in the afterlife.[15]

Quebec's more aggressive secular nationalists—long repressed under Duplessis as demonic "leftists"—seized the opportunity presented by the Jean Lesage's "Quiet Revolution" to redefine Saint Jean-Baptiste Day as a more overtly assertive expression of Quebec's distinct fate within Canada. The parade's political symbolism became ever more strident throughout the 1960s, reaching something of a culmination in the tear-gas-filled night of June 24, 1968.[16] In 1977, at a time when Lévesque's sovereigntist (*souverainiste*) Parti Québécois (PQ) government was consolidating its control over the province, the holiday became more simply known as the "Fête Nationale."[17] As some observers have noted, "The saint and his sheep were put out to pasture."[18]

The Feast of Saint Jean-Baptiste lost much of its political fervor during the 1990s, especially in increasingly cosmopolitan Montreal. Popular outdoor concerts surpassed the importance of the parade as increasing numbers of Montrealers viewed the holiday more prosaically as an opportunity to enjoy one of the year's longest days in a climate all too often dominated by ominous clouds and dark skies.[19] Even travel writers from the Canadian Rockies began to advise those in search of more grand celebrations to travel 200 miles downriver to Quebec City to fully appreciate Saint Jean.[20]

## The Accidental Prime Minister

Returning to Saint Jean-Baptiste Day in 1968, the Pierre Elliott Trudeau under fire in the stands in front of Lafontaine Park had first burst onto the Canadian political scene three years before. He was, of course, well known within Quebec for his increasingly strident antinationalist stance when discussing the issues of the day.[21] He nevertheless was a fresh face to Canadians elsewhere.

Trudeau's rejection of nationalism ran deep. In 1967, he expressed himself with singular clarity on the subject, writing, "I early realized that ideological systems are the true enemies of freedom."[22] This rejection of viewing the world through abstractions was especially pointed on the subject of nations. Five years earlier, he had written in *Cité Libre* that, when thinking of the nation, it is important "to keep three things in mind. The first," he continued, "is that the nation is not a biological reality—that is a community that springs from the very nature of man. . . . The second is that the tiny portion of history marked by the emergence of the nation-states is also the scene of the most devastating wars, the worst atrocities, and the most degrading collective hatred the world has ever seen."[23] Finally, he added "to insist that a particular nationality must have complete sovereign power is to pursue a self-destructive end."[24]

Trudeau's obstinacy on such issues eventually would become something of a trademark. Three decades later, Ontario premier Bob Rae of the New Democratic Party would observe that "in his taste for winning the argument pure and simple, Trudeau failed to seek the middle ground."[25] Back at the time of his elevation to prime minister, Trudeau was still better known outside Quebec for having been deported from Yugoslavia to Bulgaria, for having swum the Bosporus, for having been arrested in Palestine as an Israeli spy, for having snuck into China, for having thrown snowballs at Lenin statues in Moscow, and for having dated pop diva Barbara Streisand.[26]

Liberal prime minister Lester Pearson brought Trudeau, by then a no-longer-young forty-nine-year-old law professor, into the Liberal Party for the November 1965 parliamentary elections. Pearson, the model of Ontario probity, was a natural conciliator (having won the Nobel Peace Prize eight years before).[27] Pearson already had undertaken a number of initiatives to try to improve relations between French and English Canadians, including extensive celebrations of the centennial of the Canadian Confederation of 1867 focused on a World's Fair in Montreal, and the adoption of the red Maple Leaf flag as the Canadian national banner. The new flag, immediately dubbed by angry monarchists as the "Pearson pennant," eschewed the traditional red-white-and-blue colors of the British monarchy.[28]

At a more substantive level, Pearson sponsored the formation of the Royal Commission on Bilingualism and Biculturalism, which had begun its work in 1963 under the joint chairs of Davidson Dunton and André Laurendeau.[29] The policies recommended by that commission through a number of distinguished publications between 1967 and 1971 set forth the outline for a new conception of Canadian federalism.

The commissioners embraced a vision of Canada predicated both on a bilingualism in federal services, which made it possible for francophones to function in their native language everywhere in the country, and also on a multiculturalism that would defuse French-English tensions by placing them in a wider sea of ethnic and cultural difference.[30] Often associated with Trudeau, who vigorously embraced his predecessor's vision for a re-

newed Canada, such policies defined a Canadian reality that had been unthinkable before Pearson launched his assault on the Imperial British un-
derpinnings of the Dominion of Canada.

Pearson was correct to be concerned with growing discontent within
French Canada. The "Quiet Revolution" unleashed by Jean Lesage's Liberals eroded long-held beliefs and political customs. The Quebec intellectuals Charles Gagnon and Pierre Vallières founded the radical Front de la
Libération du Québec (FLQ), an organization that would unleash several
terrorist actions in Quebec and elsewhere between 1963 and 1970. In 1964,
the Quebec Liberal Party—the Fédération Libérale du Québec—split off
from Pearson's own federal Liberal Party. In 1966, Daniel Johnson led a revitalized nationalist Union Nationale Party to victory over Lesage's profed-
eral Liberals.[31] Pearson properly understood that popular support for Canadian federalism was on the run in Quebec, if it had ever existed at all.

At the prompting of his senior adviser, Maurice Lamontagne, Pearson
moved to shore up his own standing in Quebec during 1965, and in the
process he tried to strengthen the ties between that province and the re-
mainder of the country.[32] Lamontagne negotiated the candidacies of Gerard
Pelletier, Jean Marchand, and Pierre Elliott Trudeau in upcoming federal parliamentary elections. Pearson's "Three Wise Men" presented themselves as
a package, helping to secure Pearson's November 1965 victory over a Tory
party led by the curmudgeonly westerner John Diefenbaker.

Pearson later recounted that he had been most taken with Marchand,
whom he described as "tough, astute, forceful, with a charming personality. He had learned to survive in the savage jungle of trade-union leadership."[33] Pearson greatly respected Pelletier, whom he found to be a moving
orator.[34] Trudeau, who had been something of an afterthought for Pearson,
became a marginally successful parliamentary secretary.

Yet Trudeau's political star began to rise quickly. He secured the Justice
Ministry portfolio, a post that placed him in the position of debating Quebec Premier Johnson over issues of constitutional reform. In December
1967, Trudeau captured the imagination of the Canadian press and public
by declaring in Commons that "there is no place for the state in the bed-
rooms of the nation."[35]

A charming bachelor of somewhat randy reputation, Trudeau immediately became a media star often compared with the martyred American
president John F. Kennedy. Miniskirts were suddenly "in" on the campaign
trail, while Hollywood stars such as Frank Sinatra flocked to be seen with
Canada's most un-Canadian celebrity politician.

The aging Pearson stepped down in April 1968, with the Liberals hitching their political bandwagon to the flashy Trudeau.[36] Pearson noted that many among the Liberal elite—such as Paul Martin (the father of the twenty-first-century Canadian prime minister of the same name)—simply could not fathom Trudeau's appeal. "The Trudeau campaign completely bewildered the old pros," Pearson would later write.[37]

Trudeau quickly made Pearson's policies for a renewed Canadian federalism his own, and thus he found himself sitting in the reviewing stand of the 1968 Saint Jean-Baptiste Day parade as the object of scorn from Quebec sovereigntists (*souverainistes*) who, in many instances, came from backgrounds and professions rather close to his own. His victory over the Progressive Conservatives led by the decent, but charisma-less Nova Scotian Robert L. Stanfield, appeared to be almost anticlimactic following the dramatic events of the previous evening.[38]

Trudeau's presence on Saint Jean-Baptiste Day in 1968 marked another moment in time when various forces that would profoundly shape national, provincial, and metropolitan life for the next quarter-century came together. Canada, Quebec, and Montreal were about to embark on a grand adventure in which every assumption about what constituted the good society would be challenged. Within two years, now–prime minister Pierre Elliott Trudeau would place his hometown under martial law.

Simultaneously, this time would be precisely the same period when new waves of migration from outside the North Atlantic basin began to transform how everyone in Montreal would live out their lives. Many of the voters who eventually determined the final outcome of the battles so firmly engaged by 1968 had yet to arrive on Canada's shores.

## Suspending Civil Liberties

Shortly after 8 o'clock on the morning of October 5, 1970, two armed men abducted the British trade commissioner, James Cross, as he left his house for work.[39] A cell of the FLQ claimed responsibility, demanding the release of "political prisoners," $500,000 in gold, publication and broadcast of the *FLQ Manifesto,* and safe passage for themselves to Cuba or Algeria. Trudeau immediately began to articulate an increasingly tough response on behalf of the Canadian government, a position that won the ready approval of Quebec premier Robert Bourassa.

A number of prominent Quebec opinion makers—such as Claude Ryan of *Le Devoir* and René Lévesque—sought to define a middle ground, with various news outlets making all or some portion of the *FLQ Manifesto* available to their readers, listeners, and viewers.[40] On the evening of October 10, however, another FLQ cell kidnapped Pierre Laporte, the provincial minister of labor, at his house. Both FLQ groups began to issue contradictory demands, with those holding Laporte staking out an increasingly hardline position.

Troops were dispatched on October 12 to protect officials and prominent citizens in Ottawa and elsewhere. Tensions continued to grow following Trudeau's statement the next day: "Yes, well there are a lot of bleeding hearts around who just don't like to see people with helmets and guns. All I can say is, go on and bleed, but it is more important to keep law and order in the society than to be worried about weak-kneed people." Numerous leading Quebec public figures—including Lévesque, Claude Ryan, and Jacques Parizeau—spent the next day trying to convince Bourassa's provincial government to negotiate with the FLQ.

More deadlines passed and, on October 15, 1970, Prime Minister Trudeau announced the imposition of the War Measure Act, essentially suspending civil liberties throughout the Province of Quebec.[41] The FLQ was declared to be an unlawful association, with associates subject to jail time. Trudeau's actions won wide approval—except for noteworthy criticism from Ryan, Lévesque, and federal New Democratic Party leader Tommy Douglas.[42] A total of 250 arrests were made within forty-eight hours in some 1,600 raids, with 247 additional arrests to follow (435 arrestees eventually were released without charge, while only 2 of the remaining 62 detainees were ever convicted of a crime).[43] On October 17, the group holding Cross declared that they were suspending their death sentence against the British diplomat. The next day, however, the body of Pierre Laporte turned up in an isolated corner of Montreal, evidently having been killed the day before in retaliation for the invocation of the War Measures Act.[44]

Jean Drapeau was reelected as mayor with 92 percent of the vote on October 25, as the vise of the War Measures Act began to tighten.[45] Arraignments under the act continued throughout November. Montreal police eventually released James Cross on December 3, 1970, with the abductors gaining free passage to exile in Cuba. Federal troops were fully withdrawn from Quebec by January 4, 1971. Convictions of the main perpetrators of the Cross abduction were tried and sentenced a year later.[46] Various court cases dragged on into 1974 before the official file on the October 1970 events would be closed.

Montreal, Quebec, and Canadian society generally were appalled by La-porte's murder. A decade of increasingly violent attacks against Canadian federalism came to an end—with no more incidents to follow.[47] The politics of Quebec identity, independence, and Canadian federalism shifted to electoral politics. Lévesque's Parti Québécois continued to garner strength until it formed the majority in the provincial parliament—Quebec's National Assembly—following a sweeping election victory on November 15, 1976.[48] The fight over the place of Quebec in the Canadian federal system continued in one form or other for another three decades, at least until the defeat of what is at this writing the most recent PQ government in April 2003.[49]

Much of the commentary about the October Crisis focused on its national significance either for Canada or for the nation of Quebec.[50] Trudeau's decisive action defined an antinationalist vision for Quebec and Canada, which eventually became enshrined in the two most enduring accomplishments of his long public career: the Canadian Constitution of 1982, and the ratification of its accompanying Charter of Rights and Freedoms. He subsequently would come out of retirement to defend the charter in 1987 following the Meech Lake Accord, and again in his famous 1992 "Maison Egg Roll" speech against the Charlottetown proposals to reconfigure Canadian federalism.[51]

Just as the race riots that erupted in Washington following the April 1968 assassination of Martin Luther King Jr. are viewed as integral to the national tragedy of American race relations, the October Crisis is seen as a turning point in the broad story of the troubled relationship between Canada's French and English communities.[52] As in Washington, so too in Montreal, such traumatic national events are local moments in time when the divisions of troubled urban communities became more starkly drawn. The October Crisis and Trudeau's imposition of the War Measures Act traumatized Montreal for well over a generation. As in Washington, these would be precisely the same decades when local neighborhoods increasingly became home to growing numbers of arriving migrants from abroad.

## A City in Turmoil

Montreal remained a city in turmoil for years. The city stood at the epicenter of a society that was at war with itself. Trudeau's vision of universal individual rights could not coexist with Lévesque's collective cultural rights; Ottawa's federalist vision could not coexist with Quebec's sovereigntist vision; personal choice in educational values could not coexist with laws establishing the language of instruction by legislative fiat; and an image of a multicultural Canada could not coexist with a Quebec still seeing itself as the victim of a defeat that had transpired on the Plains of Abraham more than two centuries before. All these conflicting forces converged on the figure of Jacques Parizeau.

Parizeau appeared to be a jovial enough politician until the driving force in his political life arose for discussion: Quebec sovereignty.[53] Jacques was the bright and proud son of the Montreal haute bourgeoisie Young Jacques trundled off to Paris and London to receive his education. A prodigy of sorts, he graduated from the prestigious School of Political Science ("Sciences Po") in Paris before moving on to defend a doctoral dissertation in economics at the London School of Economics and Political Science. Nothing prepared him, therefore, for the indignities of speaking French in the Montreal of the 1950s. He was not accustomed to being told to "speak white" (as the more hateful defenders of the language of Shakespeare would shout at him upon hearing his French in the city's bars and on its streets).

Parizeau eventually made his way into the local banking and academic communities, falling naturally into politics as an economic adviser to Quebec Liberal premier Jean Lesage during the "Quiet Revolution" of the 1960s. As with so many Quebec intellectuals of the time, Parizeau was drawn to the opportunities created by the secularization that accompanied the interlocking and profound social, economic, and political changes promulgated by Lesage. Like so many of his companions, Parizeau nurtured a deep grudge over anglophone arrogance and domination. In 1968, he boarded a train in Montreal's Windsor Station headed to a national federal-provincial conference on behalf of Lesage, planning to use the trip to Banff in Alberta as quiet time to prepare his report. He left Montreal a federalist and arrived in Banff a sovereigntist (*souverainiste*).

Quebec's displeasure with the intricacies of federal fiscal relations within Canada were—and are—not unique to that province. Many Canadians harbor their own grudges against federal bureaucrats in Ottawa. The French reality of Quebec adds urgency to such complaints, bringing together a potent brew of regionalism and cultural identity. In the case of Parizeau, economic analysis combined with a large chip on his shoulder left over from his dark days in the Montreal of the 1950s. He served in René Lévesque's Parti Québecois sovereigntist (*souverainiste*) government as finance minister into 1980s, before resigning with flare while drinking expensive wine at his final press conference.

Parizeau returned to electoral politics to take over the foundering PQ following the defeat of Lévesque's successor, Pierre Marc Johnson. A master manipulator and an eloquent if at times bombastic speaker, Parizeau brought the party back to power. His single-minded goal had become winning a provincial referendum on the question of Quebec sovereignty.

Every moment in Parizeau's life pointed to the October 1995 referendum on Quebec sovereignty. A majority of voters in 80 of Quebec's 125 "ridings" (assembly districts) vindicated Parizeau's long quest by supporting his position.[54] Well over a majority (perhaps as much as 60 percent) of Quebec's francophones agreed with Parizeau and voted "Oui" on the sovereignty issue. Yet, as referendum night turned into the morning after, a final tally fell shy of his dreams: 50.6 percent, No; 49.4 percent, Yes. Nonfrancophones throughout the province together with residents of the Island of Montreal had managed to save Canadian federalism by the narrowest of margins.[55]

An exhausted (and perhaps inebriated) Parizeau climbed to the podium in the steamy Palais des Congrès in Montreal together with a more suave and proper Lucien Bouchard, the leader of the Bloc Québécois in the Canadian federal Parliament. Not surprisingly, Bouchard and Parizeau vowed to fight on.[56] Anger, frustration, and indiscretion overtook valor as Parizeau began to speak. "It's true we were beaten," he declared, "But by what basically?" His answer would send Parizeau packing off to political oblivion within twenty-four hours (and similarly would force the sovereignty debate off the Quebec political agenda for at least a decade, if not more).[57] "L'argent et les ethnies" ("By money and the ethnic vote"), he cried out.[58]

Ironically, Parizeau was partially correct. Nonfrancophone Quebeckers, especially the so-called allophones (immigrants and their children whose native language is neither French nor English), defeated the sovereigntist (*souverainiste*) juggernaut. Parizeau's unpardonable sin was not speaking falsehoods. Rather, it was speaking intolerance.[59]

Like so many of the world's great cities, Montreal was no longer the divided city of Parizeau's youth. Somewhere along the way, tens of thousands of migrants had arrived from around the world, creating a middle ground for themselves between the French and English, Protestant and Roman Catholic, metropolitan and rustic worlds of old Montreal and Quebec. Lodged apart from, yet within, cultural battles that they did not prompt, forced to send their children to French-language schools when many would have preferred to study in an anglophone environment, "allophones" added just enough ballots to the "anglophone" vote to defeat Parizeau's sovereigntists. For these new arrivals, Quebec was not simply the domain of the descendants of French founding families and English conquerors. Quebec was theirs as well.

The city bristled with hostility and tension, accompanied by a consequent increase in creativity and intellectual turmoil. Each store owner fined for displaying a sign in an outlawed language (e.g., English), each child enrolled in a French-language school against a parent's will, each cab driver told by an outraged anglophone to "speak white" added to a volatile mix that would culminate in the bitter October 1995 referendum on sovereignty.[60] The apparent detoxification of Montreal a short decade later provides testament to the soothing presence of thousands of new Montrealers who have arrived in the city—and on the continent—in the months and years since that time.[61]

During the thirty-five years following October 1970, four Liberal Party premiers (Robert Bourassa, twice; Daniel Johnson Jr.; and Jean Charest) exercised authority over the Province of Quebec for seventeen years and eight months; while five sovereigntist (*souverainiste*) PQ premiers (René Lévesque, Pierre-Marc Johnson, Jacques Parizeau, Lucien Bouchard, and Bernard Landry) have held sway for seventeen years and four months.[62] These swings represent far more than the give-and-take of two competing political parties. Quebec's Liberals and their foes in the PQ embodied two competing worldviews and ideological belief systems. The first embraced values predicated on individual responsibility and rights as protected by a largely benign Canadian federalism; the second engaged the rhetoric of the nation, the collective, and corporate rights endangered by continuing subjugation to Ottawa.

The initial engagement of these mutually exclusive worldviews during the years between the asbestos strike and Pierre Elliott Trudeau's emergence as Canadian prime minister left no middle ground between them. Few electoral battles have ever been fought as bitterly as Quebec's two referendums on sovereignty (the first, on May 30, 1980, ended with 40.5 percent of voters supporting Lévesque's "sovereignty association" with Canada, while 59.5 percent rejected any renegotiation of Quebec's status within Canada; the second was the above-mentioned razor-thin poll on October 30, 1995).[63] The embrace of a new Canadian Constitution over the expressed opposition of Quebec's elected leaders—at Meech Lake in April 1987, and again at Charlottetown five years later—left little room for accommodation with the rest of Canada.[64] Quebec's Pierre Trudeau (1968–79, 1980–84), Brian Mulroney (1984–93), Jean Chrétien (1993–2003), and Paul Martin (2003–) have served as Canadian prime ministers during this period of intense constitutional debate, often bringing their home province's family feud to center stage in Ottawa.[65]

## The Boss

The political, cultural, and ideological turmoil swirling through Montreal's streets, bars, classrooms, theaters, sports arenas, and airwaves somehow parted over City Hall. Jean Drapeau, who had been reelected at the height of the October Crisis, continued in office until his retirement in 1986 (by which time the legendary mayor had served just shy of thirty years as the city's senior official).[66] Drapeau—who was known for bringing the Olympics, a World's Fair, Major League Baseball, and a world-class subway system to Montreal together with enormous municipal debt—initially ran as an anti-Duplessis reformer in 1954 on his reputation as the well-known defender of the asbestos strikers. An eventual recipient of the French Légion d'Honneur, Drapeau gained international notoriety by standing at General Charles de Gaulle's side in July 1967 when the French president unleashed a torrent of emotion by exclaiming "Vive le Québec libre" from the balcony of Montreal's City Hall.

Drapeau had something of an imperious Gaullist style himself, despite his early reformist image. The mayor ran up unprecedented municipal deficits in his quest to construct a world-class infrastructure beyond the parameters of prudent public financial practices. He did so by holding a tight reign over the physical development of the city and by command and control of the municipal bureaucracy and City Council. Drapeau sought to occupy 100 percent of Montreal's municipal "political space," driving provincial and national party organizations out of city politics in the process.

Yet the mayor was not as omnipotent as his image might suggest. Opposition shifted from electoral parties to municipal unions, which were increasingly assertive. Drapeau's massive construction projects often destroyed neighborhoods—which only generated a burgeoning community-based social movement. A fledgling opposition began to take shape during the early 1970s, energized by the broader conflicts over Quebec sovereignty. By 1973, a loose coalition "of left intellectuals, community organizers, and trade unionists" formed an umbrella group—the Montreal Citizens Movement (MCM)—to contest the 1974 municipal elections.[67] They gained eighteen seats on the City Council after that initial campaign, a number that would continue to grow until the MCM had gained control of City Hall under the leadership of Jean Doré following Drapeau's retirement in 1986.[68]

Doré viewed democratic governance as an iterative process that built on community and neighborhood civic associations over time.[69] His primary objective, like that of the MCM as a whole, was to move Montreal toward a more "consultative" system of decisionmaking in which citizens and communities would have greater control over their own lives.[70]

The MCM's rise—which accompanied the fights over sovereignty and the arrival of the PQ in power in Quebec City—was concurrent with an explosion of grassroots community groups receiving financial support from the Montreal municipal and Quebec provincial coffers. As Henri Lustiger-Thaler and Eric Shragge report, "From 1973 to the end of the 1980s, the number of groups in Montreal went from 138 to 1,500 and those receiving money from the Quebec government's Ministry of Health and Social Services went from 31 to 547."[71] Doré, a low-key executive, held power until losing a hotly contested race in 1994 to Pierre Bourque.[72]

Local analysts argue convincingly that the demise of the MCM was tied to profound changes in the broader Canadian, continental, and even global urban environment. Facing the growing pressures captured in shorthand by the turn of phrase "globalization," Montreal confronted increasing competition from other cities that were developing an "entrepreneurial" style. Neither the city nor the province could sustain the massive public expenditures on which the community movement of the 1970s and 1980s had been built.[73]

Bourque, for his part, was the son of a pro-Drapeau City Council member (his father had represented Rosemont from 1957 until 1970). The younger Bourque trundled off to study engineering and horticulture in Belgium before returning to launch his career as a landscaper at the Expo 67 World's Fair.[74]

Bourque fils evidently used his political ties well, garnering an appointment as the director of the Botanical Gardens. Bourque credits retired mayor Drapeau with having encouraged his entry into electoral politics in 1994 (Drapeau died in August 1999). Bourque came to office just as the battle over the second referendum on Quebec's status within Canadian Federation reached a full boil. Bourque's successes, which were considerable despite his ultimate defeat, depended on his ability to blend the interests of Quebec nationalists with those of transnational migrants who were flowing into increasingly cosmopolitan neighborhoods all over town.[75]

## Tending the Urban Garden

Pierre Bourque had returned from Brussels a young sovereigntist drawn to the nationalist firebrand Pierre Bourgault.[76] Like many nationalists of his generation, he abandoned more radical strategies following the creation of René Lévesque's Parti Québécois in 1970. Bourque aligned himself with Jacques Parizeau as Parizeau rode Lévesque's coattails into the first PQ government in 1976.

Bourque had other preoccupations at the time. As the director of Montreal's Botanical Garden, young Pierre was busy trying to upgrade the long-ignored facility to convert it into a major municipal asset. Bourque would succeed in this task beyond his wildest imagination. His quarter-century custodianship of the garden coincided with its steady elevation into the ranks of the best botanical gardens in the world (regularly being compared with the "Big Four" in London, Berlin, New York, and Paris).[77]

His engagement in politics and management of the Botanical Garden slowly drew Bourque away from the nationalist nostrums of his youth. Bourque's vision for the garden—which occupies 75 hectares around the Olympic Stadium and attracts 1.3 million visitors a year to its rose gardens, Chinese Pavilion, and world-renowned Biodome—was consistently intercultural.[78] In the spring of 1994—just as the politics of sovereignty were reaching a new crescendo following the debacles of Meech Lake and Charlottetown—Bourque announced his intention to run for mayor. In so doing, he founded a new party—Parti Vision Montréal—that set out to defeat the unappealingly bland Jean Doré. On November 6, 1994, his new party garnered thirty-nine of the City Council's fifty-one seats.[79]

Once elected, Bourque embraced all Montrealers as his constituents as his own—even as his former supporter Parizeau was berating ethnic allophones as the sworn enemy of Quebec independence.[80] Bourque spoke with increasing passion about the need to include everyone in the city's plans and visions for the future.[81] Nevertheless, Bourque's easy working relationship with PQ governments in Quebec City—combined with his sovereigntist past—made many in the anglophone community uneasy with his leadership.

Bourque returned easily to a second term in City Hall on November 1, 1998.[82] After some initial squabbles with the Péquistes in Quebec City, the mayor slowly began to lay out his plans for an amalgamated city government embracing the entire Island of Montreal.[83] In doing so, he was pursuing an unrealized dream of his early patron Mayor Jean Drapeau.[84]

Municipalities in Canada are the administrative creatures of their provinces. Bourque thus required additional support from Quebec premier Lucien Bouchard to amalgamate all the various jurisdictions on the Island of Montreal under a single metropolitan-scale city government. Bourchard, seemingly sensing an opportunity to engulf the city's western English and immigrant suburbs within a French-dominated megacity, enthusiastically embraced Bourque's proposal.[85] Superficially, the ensuing campaign followed the well-worn script of confrontation between anglophones and francophones. Suburban mayors—often representing English-speaking constituencies to the west—vociferously opposed Bourque and Bouchard's "one island, one city" plan.[86] Bouchard and his party pushed the amalgamation plan for Montreal forward, because he had similar proposals for several other cities across the province. The newly enlarged city extending throughout the entire Island of Montreal officially came into existence on January 1, 2002.[87]

Bourque and Bouchard overplayed their hands. Quebec voters went to the polls on November 8, 2001, to elect new municipal councils for all the province's towns and cities—including the soon-to-be-created large Montreal. Angry suburban voters—both francophone and anglophone—took out their wrath on having had their municipalities abolished by electing Gérald Tremblay mayor of the new mega-Montreal. Tremblay, who had been a minister of industry under the last Bourassa Liberal Party government in Quebec City, was a lawyer with a master's in business administration from Harvard and long-standing ties to the Liberals.[88] He handily defeated Bourque, who nonetheless had carried 60 percent of voters within the old city boundaries.[89]

## A Montreal Open to the World

Tremblay's ally, the provincial Liberal leader Jean Charest, rightfully saw a wedge issue on which to build a Liberal electoral majority in the province. Charest promised that those communities that had been abolished through the mergers of the previous year would have an opportunity to repeal their fates should the Liberals party regain control of the National Assembly in Quebec. Bourque's ally, PQ premier Lucien Bouchard, found himself with a political albatross pulling support away from a party that advocated sovereignty at a time when sovereigntist (*souverainiste*) sentiment was on the wane. Almost inevitably, Bouchard's PQ went down to defeat in the April 2003 provincial elections.[90]

The story does not end quite so simply, however, for neither Tremblay nor Bourque played according to the old script of Liberals versus Péquistes. Tremblay's greatest ambition has been to stimulate Montreal's economic development. As befits a well-trained holder of a Harvard master's in business, the new mayor appreciated the managerial advantages of a larger, more inclusive city. Though anti-Bourque, Tremblay backed mild decentralization within a metropolitan-scale municipality, rather than endorsing an overtly antimerger agenda. Bourque, for his part, primarily pursued his "one island, one city" campaign as a means for gathering greater resources for municipal government (even if his PQ patrons in Quebec City viewed the amalgamation as a vehicle for submerging antisovereigntist anglophones and allophones under a francophone sea). Bourque wanted to accomplish one of Drapeau's few unrealized dreams: extending the city limits to encompass the suburbs.

Although he was one of the first victims of a municipal revolution of his own making, Bourque continued to retain electoral support within the old city boundaries.[91] He abandoned his PQ ally Bouchard just as the 2003 provincial electoral campaign began. Bourque threw his remaining popularity behind a new provincial party—Action Démocratique du Québec— in a move that stunned his old allies in the PQ.

Bourque's endorsement of one of the PQ's main rivals in the National Assembly in Quebec City did not end with merely changing parties. The former sovereigntist, who had once been associated with the likes of Bourgault and Parizeau, seized the opportunity provided by the press conference at which he declared his candidacy for the National Assembly to simultaneously renounce the prosovereigntist movement. Bourque urged citizens of Quebec to plan for a future within the Canadian Confederation and to embrace English as a means of communicating with the world at large. His manifesto—"A Quebec That Is United and Open to the World"—stunned local journalists and voters.

Bourque went beyond embracing a federalist future for Quebec. He criticized the sovereigntist policies of the past. For example, while acknowledging that Bill 101 had been necessary to consolidate a French identity in Quebec, Bourque observed that this language law had alienated nonfrancophones, provoking an unfortunate exodus out of Montreal and Quebec.[92] "If Quebec cannot succeed, by its strengths and its will, in integrating a community of 800,000, which expresses itself mostly in English and is concentrated in Montreal," he declared, "then I would not bet on the survival or the fulfillment of Quebec itself."[93] Quebec, he concluded, "must turn away from fantasies of independence and sovereignty."

Many factors contributed to Bourque's seemingly abrupt change of sentiment. As with any politician, electoral calculations surely mattered. Bourque's sense of his city's future as being tied to a larger world evidently intensified as he viewed what he saw as deepening isolationism next door in the United States following the September 11, 2001, attacks on New York and Washington. Whatever his past failures, Pierre Bourque appeared to have achieved a dream that had long eluded his predecessors: The city's boundaries now extended to cover the entirety of the Island of Montreal.

Something more was at stake as well. Pierre Bourque, as his various plans and programs at the Botanical Gardens revealed time and time again, believed in the power of intercultural diversity. His party Vision Montreal embraced Montreal's varied population. The party activist Helena Jean-Louis expressed this commitment well at a 2004 party convention. Jean-Louis, herself a member of the city's Haitian community, told a reporter from the *Montreal Gazette* that party members "belong to other cultures that come from abroad, while still being part of Quebec culture."[94]

The cosmopolitanism of his beloved Montreal could never be reconciled with the sovereigntist dreams of Quebec nationalists. The Canadian Broadcasting Corporation journalist Peter Black seemed to cut closest to the core of the issue when he wrote:

> Throughout his eight years as mayor of Montreal, Bourque worked assiduously to bring the city's many cultural communities into his Vision Montreal tent. He scarcely missed an ethnic festival or a media opportunity, including bunking out for a weekend with a poor immigrant family. Whatever drives Bourque, it's clear his life's passion is making Montreal a flourishing, exciting, and exotic place, just as he did with the Montreal Botanical Gardens where he first made his name and fame. His unbending support for the PQ plan to merge the island's municipalities into an island-wide city may have sown the seeds of his own political defeat, but it helped accomplish a dream Montreal mayors have had for decades.[95]

Bourque and the Action Démocratique du Québec ran poorly in the 2003 provincial elections. The PQ candidate Diane Lemieux defeated him in his own riding (parliamentary district). By late spring of that year, the once popular and successful mayor of Montreal Pierre Bourque was merely an opposition leader on the Montreal City Council.[96] Bourque's evolution in the years before becoming Montreal mayor, during his eight years in office and ever since, largely paralleled the political and social transformation of his city from a community sharply divided between French and English into an urban garden of multiple hues and cosmopolitan subtlety.

## Not Quite "One Island, One City"

Despite their victories, Quebec premier Charest and Montreal mayor Tremblay still had to confront the merger issue as they consolidated control over the province and city. Both leaders had won votes by promising to revisit the merger of cities throughout the province with their suburban neighbors, including combining all of the municipalities on the Island of Montreal into a single megacity.[97] Due to the traditionally strong hand of Montreal mayors such as Drapeau and Bourque, Charest and Tremblay attempted to redefine regional consolidation as a means for *decentralizing* the government and expanding the role of neighborhoods in municipal governance. As one of Canada's leading specialists on municipal administration, Andrew Sancton, observed:

> Legislation amalgamating the municipalities on the Island of Montreal provided that borough councils would replace suburban municipalities and, more importantly, that nine such councils would be established within the territory of the old city. Considering that borough councils were given direct control over local zoning decisions and a surprisingly extensive list of local services, the main effect of amalgamation within the old city was, paradoxically, a remarkable degree of political decentralization.[98]

Tremblay's early proposals would have granted greater taxation powers to the boroughs, expanding the role of borough council chairs to that of local "mayors."[99] Consequently, Sancton argued, the "key point about governance debates in Montreal since amalgamation is that they have focused on weakening the control of the centre."[100]

In other words, Montreal and Quebec politicians had stepped through the looking-glass when they amalgamated all the jurisdictions on the Island of Montreal into a single municipality. Battle lines were drawn initially between the then-sovereigntist government in Quebec City and predominantly anglophone suburban communities, between inner-city neighborhoods and City Hall, between those interested in the "rationalization" of public services and those concerned with transparent governance in their city.

Tremblay repackaged the question as he moved through his election campaign and began his administration.[101] He and Charest quietly reformulated taxation and service codes in such a way as to ensure that money would continue to flow from demerged suburbs to regional service authorities. Consequently, as bemused commentators in Toronto's *Globe and Mail* pointed out with a more than lagniappe of glee, suburbanites who voted for demerger were endorsing a plan that disenfranchised them from participating in many of the decisions concerning the allocations of their own taxes and services.[102]

Moving ahead on their own, Tremblay and Charest reconfigured local politics in such a way as to render the 1970s and 1980s debates over Quebec sovereignty increasingly irrelevant to municipal concerns during the early decades of the twenty-first century.[103] Montreal politics, like the city itself, was undergoing substantial change.

Former mayor Bourque, sensing that the fault lines running through the amalgamation debate were shifting from older patterns pitting francophones against anglophones, tried to cast the debate back into the parameters of the linguistic politics of the past.[104] Reporters from English Canada similarly tended to read historic disputes over language into more contemporary disputes over metropolitan government in Montreal.[105]

As summer turned into autumn, however, Tremblay's maneuvering, general political debates, and regional opinion polls began to shift attention toward other issues.[106] Though francophones continued to view the question through different lenses than their anglophone neighbors—and were more likely to oppose demerger than anglophones—the startling fact was that clear majorities of both groups who lived within the old city boundaries endorsed the larger city, while their suburban counterparts often did not. Allophones throughout the region were largely for demerger at this time.[107]

Leading politicians within the City of Montreal and Province of Quebec jockeyed for advantage throughout the fall and winter.[108] As they did, the terms of reference within the debate began to shift in response to events on the ground. A widespread perception took root throughout the city that the new mega-Montreal administration was providing lower-quality municipal services at higher cost. The city's failure to deal effectively with unusually heavy snows during the 2003–4 winter seemed to confirm what merger opponents had been claiming for some time: The provision of public services was becoming less efficient and more expensive as a consequence of the merger.

Resentments over the manner in which the merger occurred—which was increasingly viewed as having been imposed on communities across the Island of Montreal by politicians and technocrats in Quebec City and in Montreal's Hôtel de Ville—amplified frustrations over service delivery.[109] For many in the anglophone and francophone communities alike, the new metropolitan city of Montreal was becoming the embodiment of heavy-handed, bureaucratic rule.[110]

The evolving parameters of debate exposed many of the competing economic interests that stand at the heart of Montreal's twenty-first-century realities.[111] With more and more francophones living in wealthy suburbs, the metropolitan region's language map no longer aligned neatly with its map of political disputation. Wealthy suburbanites were more often opposed to being tied together with poorer neighborhoods in the old city limits; corporate interests promoting economic development favored merger as a means of spreading tax burdens; the city's traditionally left-wing community activists were increasingly drawn to Tremblay's ingenious linkage of expanded neighborhood power to metropolitan governance. Ivor and Margaret Watkins, two demerger supporters from Pointe Claire, expressed well a certain suburban disdain for scraggly urbanites as they declared before reporters from the *Montreal Gazette* that "we moved away from Montreal in 1977 because we couldn't stand Montreal. There's no way we're going back to Montreal."[112]

By mid-2004, Tremblay and Charest had seemingly gained momentum in their campaigns to save metropolitan governance.[113] Charest managed to impose ever higher thresholds for demerger in laws and regulations governing referendums on the subject both in Montreal and in other towns and cities throughout the province that similarly had merged with their neighbors.[114] By doing so, he made the task of the *défusionnistes* ever more difficult.[115] Tremblay initially played his hand with increasing skill as well.[116] Much to the dismay of his opponents, the mayor mobilized public funds behind defeating the demerger movement.[117]

Both Charest and Tremblay worked to ensure that there could be no return to the status quo ante even if various communities approved demerger.[118] All communities on the Island of Montreal, they argued, would continue to share the costs for basic services no matter the outcome of the vote.[119] They similarly garnered support in their efforts to save metropolitan government from erstwhile enemies such as Pierre Bourque.[120]

More dispassionate observers began to note that the city's problems were no longer concentrated in a few inner-city neighborhoods. A financial analysis revealed that at least some suburban jurisdictions would be

better served by remaining in the larger megacity.[121] Studies showed that
fires were causing fewer personal injuries and less property loss following
the municipal mergers as more firefighters had become available to respond to each alarm.[122] Moreover, reports released in early 2004 indicated
that poor people were beginning to move across jurisdictional lines into
the suburbs.[123]

By mid-May 2004, demerger proponents in twenty-two of the twenty-
eight municipalities that had been merged to form the new metropolitan city
of Montreal—covering more than half of the region's population—successfully filed petitions to bring the issue to a vote in late June.[124] Elsewhere
in the province, sixty-seven communities petitioned for relief against mergers, notably around Quebec City, where the subsequent debates were particularly fractious.[125]

The final stage of the demerger campaign was launched by emotional television debates between Montreal mayor Gerald Tremblay and former
Westmount mayor Peter Trent, who faced one another on both English- and
French-language broadcasts.[126] Tremblay's quick challenges to antimerger
assertions by Trent during the second, French-language debate in particular appeared to have energized those in favor of municipal consolidation.[127]

The second debate proved to be especially contentious, because it played
out against reports that advanced voting a few days before had been especially heavy (preballoting being used in the place of absentee ballots for
those who cannot get to the polls on election day).[128] The high turnout in
municipalities hostile to amalgamation was taken as a sign of major gains
for the pro-demerger forces in the days leading up to the final voting.[129]

Editorial writers for all the province's major daily newspapers urged
their readers to support the larger amalgamated cities.[130] On June 20, 2004,
voters went to the polls in eighty-nine municipalities across Quebec—including twenty-two on Montreal Island—to decide the fate of the province's
amalgamated "megacities."[131] Once the polls closed and the votes were
counted, residents in fifteen Montreal suburbs had chosen to leave the city
behind.[132]

Leading proponents of demerger, such as former Westmount mayor
Peter Trent, claimed a rousing victory, declaring that the residents of his
wealthy English enclave had "gotten our city back."[133] Trent would soon
leave politics all together, declaring that "my job is done."[134] PQ leaders,
smarting over the collapse of one of their major policy achievements just a
few years before, urged calm while intimating that language tensions could

well increase across Montreal Island in the wake of the victory of the likes of Trent.[135]

Once analysts had more time to consider the outcome of the elections, the day's results appeared much more complex. Linguistic cleavages could not explain province-wide results in which francophone and anglophone voters accepted or rejected demerger at a more or less equivalent rate. Moreover, one-third of all communities given an opportunity to vote for demerger across the province did so, including a number of predominantly francophone suburbs around Quebec City. On Montreal Island itself, the industrial and historically independent-minded francophone town of Montréal-Est endorsed demerger, as did most traditionally anglophone suburbs to the west.[136]

Resentment over the high-handed manner in which the initial mergers had been handled motivated many who went to the polls. Voters interviewed as they left their polling places frequently echoed the sentiments of Westmount francophone Lucien Savard when he declared, "Its not about language. . . . It's about Democracy."[137] Montreal city and Quebec provincial politicians did well to read and to remember the lead headline in the *Montreal Gazette* as election day approached: "A Rude Awakening."[138]

Increasingly, with the passage of time, class emerged as a more compelling factor in the demerger vote. A strong positive correlation emerged between support for demerger and the rate of home ownership in any given community.[139] Demerger also became a debate about complex and often negative attitudes among suburbanites toward intercultural diversity in central Montreal. The city's transnational migrant and nontraditional population implicitly became an unspoken issue in how the demerger vote played itself out. Suburban disdain for the city, wealthy condescension for the poor, and middle-class angst over tottering social status—rather than linguistic animosity—drove opponents of amalgamation to vote against sharing postal addresses and public services with those less visibly fortunate than themselves.

Mayor Tremblay's opponents—such as Pierre Bourque—were quick to point out that the departure of more than half the jurisdictions incorporated into the new megacity must be considered a defeat for the mayor. The city nonetheless entered the twenty-first century in a stronger position than it had left the twentieth.[140] The clock was not turned back to where it had been before Bourque had initiated the merger/demerger process.

As Tremblay and Charest quickly asserted, 87 percent of the island's population (1,600,000 of 1,800,000 residents) will live in the new City of

Montreal once demergers have been completed by January 1, 2006—as opposed to just 55 percent when the process began.[141] More significantly, 80 percent of the island's property-tax base will remain located within city boundaries.[142] As a consequence of Tremblay's and Charest's deft maneuvering the previous year, the island's demerged municipalities must renegotiate with the City of Montreal for many communal services, with rates undoubtedly being set to the benefit of those who are living within city limits.[143] The considerable costs of the demerger process will only be added onto the suburbanites' tab as well.[144]

Language issues haunted voting booths on June 20, 2004, to be sure. Many PQ politicians and local journalists correctly detected a return of linguistic paranoia among some anglophones who embraced demerger. More striking, however, was the remarkable display of cultural and class bloody-mindedness evident among many suburban anglophones as they went to the polls. West island suburban anglophone communities dealt themselves a losing hand in the politics of municipal service provision by turning their backs on the larger city of Montreal. Blinded, perhaps, by their own anger, suburbanites gave up what little influence they may have had over the decisions made by City Hall.

The demerger process proved to be complex and expensive. The changes in legal status enacted during the demerger campaign required the city and the fifteen *défusionnistes* communities to create new financial relationships that remain far more interconnected than anything that existed before 2002. Moreover, the City of Montreal will oversee the demerger process until the departing municipalities that have voted to leave elect their own governments in November 2005.[145]

The early stages of demerger were accompanied by difficult negotiations with the city's blue-collar workers. Municipal unions and city officials such as Executive Committee president Frank Zampino and Mayor Tremblay eventually arrived at new contracts that would be binding on the city following demerger after protracted and brutal bargaining.[146] As a result of these new circumstances, the city passed its largest budget ever for 2005.[147]

Tremblay drew on his increasing popularity in those neighborhoods of the city that would remain within Montreal to stabilize city services and finances while realigning electoral regulations to favor his own supporters.[148] The new city budget lowered taxes for many in the city—especially for those Montrealers who had opposed demerger.[149] Taxes rose imperceptibly in Côte-des-Neiges and Notre-Dame-de-Grâce, for example, while surging by more than 8 percent in *défusionnistes* Westmount.[150] Residents of communities slated to leave the city began to complain vociferously about declining city services.[151]

Tremblay and his team led by Zampino, Bourque, and a new "Project Montreal" party made up of environmental activists and other core groups within the old Jean Doré coalition all sought out support among the city's increasingly diverse transnational migrant community.[152] Tremblay and Bourque eventually squared off against one another in the November 2005 mayoral elections.[153] The demerger of wealthy, and in some instances anglophone, suburbs necessarily means that allophone voters will be of greater importance after the city's boundaries are realigned once more in January 2006.

On June 20, 2004, Montreal's traditional anglophone west island communities chose to abandon city politics to their francophone and allophone neighbors. Future municipal politicians scrambled for the support of native Québécois and transnational migrant city residents. The ancient Montreal game of "French versus English" could well dissipate, at least with regard to municipal politics within the new City of Montreal.

The bitter merger/demerger fight exposed the contours of a new Montreal that had taken shape underneath the cover of the great debates over Quebec sovereignty at the end of the twentieth century. Mayor Tremblay in particular turned his predecessor's older vision, which in fact had pitted francophones against anglophones, into a twenty-first-century fight over the nature of economic development and community democracy.

Transnational migrants stand at the core of this newer Montreal. The city's economy can no longer grow without migrants, nor can local neighborhoods remain viable. The city's recent arrivals form the foundation on which a new city has taken shape—and around which the contest of local politics will be fought.

## Saint Patrick Rules

Nearly 1,900 weeks had passed since Pierre Elliott Trudeau sat on the reviewing stand outside LaFontaine Park on a long June evening staring straight forward as threatening pandemonium broke out all around him. On March 14, 2004, another new Canadian prime minister—like Trudeau all those years before, one as yet designated by the Liberal Party caucus and still awaiting the mandate of an electoral victory—sat watching a Montreal parade pass by a reviewing stand.[154] This prime minister—Paul Martin

Jr.—had waited his entire life to become the leader of his country (in part to avenge his father's defeat at the hands of Pierre Trudeau for leadership of the Liberal Party).[155]

Seasonably gray clouds hung low over the city that day, keeping the temperature just below the freezing mark. The throngs of Montrealers swirling all around did not seem to notice. Their city was a different place than Trudeau's city had been just two generations before. A new patron saint—Saint Patrick—embraced Montreal's new vibrant diversity without a hint of the stern nationalism personified by the curly haired peasant boy Saint Jean.

Martin's Canada was different from Trudeau's as well—in part because of the new patriated Constitution and Charter of Rights that Trudeau himself had put into place through force of will. The Province of Quebec was different—with the question of sovereignty having been laid to rest for a while following the Liberal Party's victory in provincial elections the previous spring.[156]

Montreal, Quebec, and Canada had become different not so much because individual Canadians had changed (although that also happened). Rather, entirely different peoples now lived in all three overlapping—yet distinct—locales. The same years that had passed since Trudeau challenged his tormentors had witnessed unprecedented migration into Canada—both in scale and in composition. Tens of thousands of present-day Montrealers never knew the embittered battles of the 1960s, 1970s, 1980s, and even 1990s—and seemingly did not care, even if they did know. Saint Patrick's Day had become theirs to celebrate a new home's intercultural diversity—leaving Saint Jean far behind.

Montrealers like to brag that their city's Saint Patrick's Day parade is older—180 years old in 2004—and larger than any of the other grand parades held in honor of that day. As many as 5,000 people may line up for Tokyo's Saint Patrick's Day parade; Chicago may dye its river green; and New York may proclaim a typically narcissistic vision of its own parade as the only one that matters. But Montreal's celebration is the granddaddy of them all.[157]

Although sufficient Irish migrants had already found a home in the city to hold a parade in 1824, the mass exodus of the Emerald Isle forced by the potato famine of the 1840s brought thousands and thousands more Irish immigrants to Montreal. Many Catholic Irish mingled comfortably with native French-Canadians (which helps to explain the presence of such prominent Quebec politicians as Daniel Johnson and Pierre-Marc Johnson a century and a half later). Both the French Canadians and the Irish Catholics, after all, shared confessional rites and similar views of British Imperialism.[158]

Irish Nationalist Thomas D'Arcy McGee—one of the leaders of the 1848 Rebellion against the British Crown—eventually made his way to Canada (via the United States). Montreal's "most popular and well-known Irishman of his day," McGee began to foresee a cosmopolitan Canada in which national, linguistic, and religious differences would matter little. Irish nationalist assassins from the United States gunned McGee down in Ottawa on April 7, 1868.[159]

McGee's vision of Canadian nationhood and Montreal cosmopolitanism is omnipresent every March on the streets of his adopted hometown. Upward of 650,000 show up each spring to celebrate Ireland's patron saint, even as fewer and fewer Montrealers bother to turn out for Saint-Jean's annual procession in far more temperate weather.[160] French Canadians, English Canadians, Catholics, Protestants, Jews, Muslims, Latin Americans, Africans, Vietnamese, and Native Americans have transformed March 17 into a celebration of a cosmopolitan and diverse Montreal created from within the social and political trauma of the late twentieth century by thousands of new city residents born elsewhere.

The United Irish Societies of Montreal—the annual parade's organizers—are proud of their holiday's new role as a symbol of diversity and tolerance of difference. Significantly, the 180th Montreal Saint Patrick's Day Parade in 2004 was presided over for the first time by a queen of color.

Tara Hecksher, a twenty-one-year-old Montrealer with an Irish dad and a Nigerian mom, grew up in Nigeria before moving to London and coming to Montreal as a student at McGill University.[161] Hecksher was the parade's first black queen, although there already had been Irish-Italian, Irish–French Canadian, and Irish-Inuit queens in the past. As the United Irish Societies spokesperson Jolyon Ditton told reporters when announcing Hecksher's coronation, "I think this is great. Montreal is a mixed cultural society . . . and I think this is nothing but good for the community."[162]

The dramatic evolution of Montreal politics from 1968 until 2004 reflected deep social, economic, and cultural changes that were taking place in the city. That transformation, in turn, was becoming institutionalized in new political arrangements and practices. In some instances, diversity capital was being created by the acceptance of new norms of appropriate behavior—as was evident in the uproar over Jacques Parizeau's intemperance following the October 1995 referendum on Quebec sovereignty. In other cases, the institutionalization of new capacities for accommodating diversity may be seen in the restructuring of local government between 2001 and 2006.

Profound changes in how Montreal functioned reveal the extent to which transnational migrants can alter the social and economic contexts within which local political processes play themselves out. Despite all the tensions and conflicts of the past forty years, the story of Montreal's social, economic, cultural, and political transformations reveals how metropolitan communities may create a new capacity for accommodating diversity in even the most seemingly hostile environment.

The 2004 Montreal Saint Patrick's Day Parade took place on Sunday, March 14, rather than on March 17, to minimize disruption to the city's normal weekday life. Anyone wandering that day into Andrew's on Guy Street, or Brutopia on Crescent Street, or the Cock'n'Bull on Rue Sainte-Catherine, or O'Hara's on University Avenue automatically would have become Irish for an hour or two.[163] They could have joined Prime Minister Martin in waving back at Nigerian-Irish-Canadian Queen-for-the-Day Tara Hecksher; and they would all have become Montrealers—not just for an hour or a day, but for life.

# Chapter 5

# Regime Change in Washington

It's a mass of irony
for all the world to see,
it's the nation's capital,
it's Washington, D.C.

—Gil Scott-Heron, "Washington, D.C.," 1994

"His public concerns were rooted in a deep faith and a commitment to a biblical vision of justice," declared the Reverend Lynn Bergfalk to a packed house at the Calvary Baptist Church on an evening in March 1997.[1] Bergfalk was eulogizing a local cyclist who had just died of cancer at fifty-three years of age. Forever dressed in ill-fitting lawyer clothes, the mysterious cyclist always had seemed to know people—especially poor people and black people. Folks living in the District of Columbia from the mid-1960s until the mid-1990s invariably encountered this apparition of a gangly, awkward giant-of-a-white-man peddling around town on what seemed to be a battered undersized bicycle. On spotting an injustice, the cyclist could easily fly into a rage.

In just a few days, mourners would file past the cyclist's metallic blue casket, which had been carefully placed behind a black-and-red ten-speed

The author acknowledges the considerable research effort of Sapna Desai, a Wilson Center intern, during the gathering of material for this chapter. The epigraph that opens this chapter is excerpted from "Washington, D.C.," by Gil Scott-Heron, published by White Metal Music, Ltd., 1994. This song is available on Gil Scott-Heron, *Minister of Information,* compact disc, Peak Top Records, 1994.

113

bicycle "adorned with roses, black bows, and the helmet of a fallen rider" in the ornate lobby of downtown Washington's John A. Wilson Building.[2] That building, which sits at 13½ Street and Pennsylvania Avenue, Northwest, within a long golf shot of the White House, had been built to house the congressionally appointed commissioners who ruled over the Nation's Capital a century earlier at the height of official Washington's Beaux Arts obsession. One might have mistaken this "District Building" for "city hall." Such a misperception would have lost sight of the fact there had been no "city" in any meaningful legal or political sense of the term ever since the 1878 "Organic Act" had reestablished congressional authority over the District of Columbia.[3]

The District Building, now known as the John A. Wilson Building, became the command central for a new local administration dominated by a young "home rule" government during the 1970s. The deceased bike rider had been a pivotal player in the municipal affairs of that era, exerting an authoritative presence throughout the building. The rider was himself tied to the building's namesake, former D.C. Council chair John Wilson, a popular local politician who had committed suicide in May 1993. The bike rider now laying in state had replaced Wilson as council chair. His name was David A. Clarke.

Clarke's family and council colleagues—together with Mayor Marion Barry Jr., who was in what at the time seemed to be the throes of the final dissipation of his political career—stood by the casket for some five hours before it was transferred to Calvary Baptist Church on Eighth Street, Northwest, for an emotional funeral service.[4] No other white man had touched the soul of black Washington quite as deeply as Clarke. On the day of his death, Ward One council member Frank Smith Jr. (an African-American) told a reporter, "I think most people didn't even know Dave was white. He probably didn't either."[5]

Dave Clarke had grown up poor in Washington in a social network dominated by African-Americans. He graduated from Howard University Law School, putting his legal skills to use in the 1960s civil rights movement down South.[6] For Clarke, and many other local activists of the era, bringing home rule to the District of Columbia was as essential an element in the civil rights agenda as were voting rights in Mississippi.

Clarke used what power he had to help poor people, as many of the mourners that day would recall. In the words of a former council legislative aide, Brigid Quinn, Clarke and his companions turned the District Building of the early 1970s into "a place that seemed as much a movement's headquarters as a government epicenter."[7] He embraced the city's new transnational migrant groups, reaching out to help his Latino neighbors during riots which rocked his long-time neighborhood of Mount Pleasant in 1991.[8]

Marion Barry, Clarke's old comrade-in-arms, captured the council chair's character at the time of his death. Recalling a civil rights meeting at New Bethel Baptist Church in the city's Shaw neighborhood during the 1960s, Barry would note, "There we all were, wearing our dashikis, had the Afros, talking for hours about black power. And there was Dave, kind of awkward in his movements, clothes never fit, but you could tell he was intensely committed to the cause. And he did it with ease."[9] Barry continued on to observe that Clarke was, at times, more popular with African-American voters than with the city's white residents.

Clarke was hardly a model legislator. Council members complained vociferously of a growing lack of direction under the irascible Clarke's chairmanship.[10] Known for his angry outbursts, he could turn any legislative session into an uncontrollable family feud.

Clarke had never been more angry than during the final months of his life, when he helped to preside over the tearing apart of a civil rights–based vision for his hometown by federal authorities. Barry recorded that, for Clarke, the transition taking place in the city "was difficult, it hurt. I remember him saying, 'Marion, I feel like I've lost my identity. People only know me now as Mr. Chairman.' He felt like he wasn't identified with a cause anymore, with being a champion of the poor or the downtrodden."[11] Clarke's last achievement was to stave off closure of the University of the District of Columbia's Law School, which would be fittingly named in his honor following his death.

Clarke's service became a funeral for an era, a moment in history when the city was run by men and women who had established their political credentials in the civil rights battlefields of the late 1950s and early 1960s. This period in the city's history was about to end.

## Washington's Civil Rights Municipal Regime

Dave Clarke, John Wilson, Marion Barry Jr., the activist and gadfly Julius Hobson Sr., the Reverend Walter Fauntroy, onetime D.C. Council chair Sterling Tucker, and many more among the city's first home rule politicians had been allies in a long fight to bring democratic local government to Washington. Their tactics were meant to shock, as when Hobson began his (in)famous Saturday rat-catching rallies, after which he gathered his prey in cages on the roof of a station wagon—the "Rat Wagon"—and threatened to release his catch in snooty Georgetown.[12]

Being tied to the civil rights movement of the era, the local advocates of home rule enlisted the support of President Lyndon Baines Johnson to seek passage of a bill transferring limited power to the District of Columbia. They did so over the staunch objections of Southern Democratic Party segregationists such as John McMillan, a powerful South Carolina member of the House of Representatives. McMillan's notorious racist response to home rule legislation was to dispatch a truck full of watermelons to the District Building.[13]

Even the masterful legislative magician LBJ failed to bring the issue to closure in Congress. Some halfway measures managed to make their way into law: The city won the right to vote in presidential elections in 1961; and it elected a nonvoting delegate to Congress—Fauntroy—a decade later, in what was "the first true citywide election for an individual public official since 1874."[14] An eleven-member elected School Board began to function in 1969, offering another proving ground for civil rights activists such as Barry to try their hand at elected politics.

The final push for home rule came in the wake of some of the worst race riots in American history, which had swept through Washington following the assassination of Martin Luther King Jr. in April 1968.[15] Once the smoke cleared, twelve people had died, an estimated $15 million in property had been destroyed, 6,300 rioters had been placed under arrest (1,660 for felonies), and the confidence of the city's white liberal establishment that it could contain black anger had been shattered.[16] Large swaths of once-vibrant African-American neighborhoods—especially along the Seventh Street and Fourteenth Street, Northwest, commercial corridors—had been destroyed, only to remain vacant for the next three decades.

Young local radicals had tried to intervene with the crowds on the streets to minimize the damage done to black neighborhoods. By the time order had been restored, a young chemist from the heart of "Blues Country"—Marion Barry Jr.—had emerged as perhaps the most visible and effective leader on the streets.

Barry was born in Itta Bena, Mississippi, and grew up in Memphis, graduating from the local LeMoyne College.[17] Following a stint in the U.S. Army, he returned to school to pursue a Ph.D. in chemistry at Fisk University in Nashville. He soon found himself swept up in the civil rights movement, dropping out of graduate school just shy of completing his doctoral studies. By 1960, he had caught the paranoid attention of the Federal Bureau of Investigation (FBI). He subsequently helped to launch the Student Non-Violent Coordinating Committee (SNCC), moving to Washington in June 1965 to head up the local SNCC office. Barry stayed on in the city as the head of Pride, Incorporated, a local self-help organization for the city's blacks.

Barry, Clarke, Tucker, Wilson, and Hobson had been major figures in launching of the "Free D.C. Movement" of the mid-1960s. Tempered by the 1968 riots, they naturally gravitated toward local politics. Local activists grew in prominence as the push for home rule gained increasing urgency following the riots. They would burst forth to gain control of the D.C. political scene as soon as hard-line segregationists in Congress—such as McMillan—could be pushed aside.

McMillan's defeat finally came in a 1972 Democratic Party primary—an election campaign during which Fauntroy, Tucker, and other local Washington leaders traveled to South Carolina to mobilize local black voters against their longtime nemesis. McMillan's departure opened the way for the passage of the D.C. Home Rule Bill in October 1973.[18] The Republican House of Representatives minority leader, Gerald Ford, led a last-ditch effort to kill the bill on the floor of the House. Ford was pulled away in mid-debate to travel across town to the White House, where he learned that President Richard Nixon would nominate him to replace his disgraced vice president, Spiro Agnew.[19] The Home Rule Bill passed, and it was signed by Nixon on December 24, 1973, and ratified by local voters on May 7, 1974.[20]

Racist Southern Democrats and conservative congressional Republicans managed to leave their indelible mark on the final legislation that granted home rule to the District of Columbia. Congress retained the right to have final authority over the city's finances, as well as effective veto authority over local laws. Unlike any other American city of metropolitan scale, Washington would be unable to tax commuters from neighboring states. Finally, the city's court and penal systems would remain in the hands of the federal government. Home rule, as granted to Washington in 1974, proved to be a nearly fatally flawed legislative compromise that has left local officials constantly running up a down escalator ever since.

In the enthusiasm of the moment, such fine points were initially lost on young community activists such as Dave Clarke, John Wilson, Marion Barry Jr., and Sterling Tucker (though not on the wily Julius Hobson Sr., who immediately understood that yet another racially motivated political swindle had been perpetrated against the city). The city's distinguished

elder statesman Walter Washington, who had been serving as the city's fed-
erally appointed mayor, retained his office in the city's first municipal elections. "Young Turks" such as Clarke, Barry, Wilson, and Tucker rode onto
the newly constituted District of Columbia Council. FBI agents assigned to
maintain surveillance on the city's leaders filed reports—the very existence
of which being rather telling in and of itself—dryly observing that "Negro
militants" were becoming elected officials.[21] Whatever divisions, rivalries,
and jealousies among them, these civil rights–era politicians would dominate local politics until Dave Clarke's death a quarter-century later.

Washington's civil rights municipal regime had been unraveling well be-
fore several hundred mourners filed past Clarke's casket. President Bill
Clinton signed Public Law 108-4 in 1995, creating a federal "Control
Board" to oversee the city's finances and management (the D.C. Financial
Responsibility and Management Assistance Authority would monitor the
District of Columbia government until 2001).[22] Clarke and his council colleagues had spent the months before the onset of his illness dismantling
many of the social institutions and programs for which they had fought so
hard at the outset of their careers.[23]

A new group of tough-minded politicians were gathering in the city's
neighborhoods waiting for the opportunity to take their place at the table of
local power. A pragmatic municipal regime was about to be born out of the
wreckage of Marion Barry's fourth administration. Barry's 1998 departure,
combined with the election of the city's first majority white D.C. Council
that year, closed the door on the city that Clarke and his colleagues had
fought so hard to bring into being.[24]


## Regime Collapse

Washington's civil rights municipal regime was increasingly in tatters by
the time of Marion Barry's third term in the late 1980s. His 1990 arrest in
a local hotel room on charges of smoking crack cocaine appeared to be a final blow to the dream of the city's civil-rights-generation leadership to
transform Washington into a showcase for their values and policies. The city
was bankrupt—financially by the flaws of the original home rule legislation; and spiritually by the degradation of ethical standards and plummet-
ing levels of professionalism within an increasingly bloated and ineffectual
city bureaucracy.

A brief moment of hope between Sharon Pratt Kelly's 1990 election as mayor and her administration's first months in office imploded in a cloud of ineptitude, unprofessionalism, ignorance, and incompetence (with a scent of arrogance tossed in for good measure). The mayor's inability to find common ground with Washington's transnational migrant residents only exacerbated her problems, as became evident in that riots that followed the arrest of Daniel Enrique Gomez on May 5, 1991.

By all accounts, Gomez was a hard-working, quiet El Salvadoran migrant who had come to Washington in 1984 to earn money to support his parents back home. By the time of his arrest, he had established himself as a serious and quiet blue-collar worker. Life was looking up as he had landed a coveted work permit, a steady job, a house-share with six other Salvadorans at Fourteenth and W Streets, Northwest, in Washington, and was learning English.[25] The fact that people liked Gomez as much as they did may well have contributed to his downfall.

On a warm Sunday evening, he and a half-dozen or so friends were enjoying the end of a soft early summer Washington weekend by tossing back a few beers.[26] Gomez, who was thirty years of age at the time, approached the Cuban owner of Don Juan's restaurant, Juan Llerena, at about 7:30 in the evening to see if he and his buddies could sit at one of the tiny restaurant's tables to finish their brews. Llerena refused. Moments later, Gomez was shot by a D.C. police officer—an inexperienced rookie cop, Angela Jewell, who with her partner had approached Gomez and his friends about violating laws against the public consumption of alcohol. Neither Jewell nor her partner spoke Spanish. According to police reports, Gomez resisted arrest, drawing a knife on the officers; according to more than a dozen eyewitnesses, Gomez was at least partially handcuffed at the time when Officer Jewell fired her gun. Hundreds of youths tossed rocks and bottles and set police cars on fire in a melee that would last nearly three nights.[27]

Gomez was hardly the only casualty of these events. An area ringed by S Street, Northwest; Twelfth Street, Northwest; Quincy Street, Northwest; Piney Branch Parkway; and Rock Creek Parkway was cordoned off, shutting down for three days one of the city's major commercial districts encompassing the lively neighborhoods of Adams Morgan and Mount Pleasant. Dozens of businesses were burned out or otherwise damaged. The newly elected mayor, Sharon Pratt Dixon (later Kelly), never recovered her tattered reputation following reports of police ineptitude that were traced back to the mayor and her coterie of top advisers.[28]

Police, court officials, and journalists failed to piece together a coherent chain of events surrounding the police shooting of Gomez. Many Washingtonians nonetheless came to realize that their city had become a different one from which the police, politicians, journalists, and many residents had thought it to be. The African-American population had recently entered what would become a sustained period of decline in older neighborhoods nestled just north of downtown and immediately east of Rock Creek Park. Both Latinos and non-Hispanic whites were moving in, with the latter significantly increasing the area's aggregate income levels.[29] Public drinking had long been a source of tension between Latino migrants and middle-class residents, two groups that were beginning to inhabit Mount Pleasant (just such a formerly African-American neighborhood).[30] The police were caught in the middle, answering calls from upscale residents to confront perturbed Central Americans who wanted little more than to enjoy life out-of-doors.

The Metropolitan Police—indeed, the entire D.C. government—had pretty much ignored a growing Latino community throughout much of the long term of office of mayor Marion Barry. By the time of the Mount Pleasant riots, more than 30,000 Hispanics had officially moved into the city, an 85 percent increase over the previous decade.[31] According to some estimates at the time, more than 60,000 Hispanics were living in the city, largely along the Sixteenth Street, Northwest, corridor focused around the riot area in Adams Morgan and Mount Pleasant.

The Mount Pleasant riots brought momentary attention to Washington's Latino community. Latino residents of the city, under the leadership of young and energetic organizers such as Pedro Aviles, attempted to leverage their newfound visibility into a sustained movement drawing on some of the same civil rights tactics used two decades before by Washington's rising African-American leaders.[32] Neighborhood leaders formed the Latino Civil Rights Task Force to investigate the underlying causes for the 1991 disturbances. That group, a year later, organized a large march from Mount Pleasant to the District Building downtown to commemorate the first anniversary of Daniel Enrique Gomez's confrontation with local police.[33]

The May 1992 protest explicitly drew on the experience of the city's Latino community with highly successful festivals. Beginning in 1970, the community celebrated its presence in the city with a popular street fair and parade held each autumn.[34] The annual event marked one of the few times when Washington's diverse Latino community came together in unison. Migrants from the Caribbean and from Central and South America alike embraced the festival, thereby creating a new sense of singular identity from among many diffuse groups.[35] Community leaders sought to draw on an emerging pan-Latino identity that transcended ethnicity, class, and regional differences to enhance the Latino political presence in the city. In this way, they were attempting both to advance a program for change and also to defuse some of the pent-up anger so powerfully released following the Gomez shooting.[36]

The protests over the mistreatment of Gomez, a growing list of complaints about the absence of concern for Latino issues among city officials, and the success of the Latino Festival would contribute to the slow but steady formation of a transnational migrant agenda in Washington city politics over the next decade and a half. The collective impact of these actions proved less than immediate. The ineffectual Kelly administration was replaced by an even less capable fourth Barry administration. The city eventually landed under the control of the congressionally appointed the Financial Control Board mentioned above. Because it was by now broke, the city had few funds to address Latino concerns even had there been the political will to do so. Increasing deficits, ever more hostile congressional oversight, and the spectacle of its longtime mayor being held in federal detention in Richmond all undermined the city's capacity to manage itself.

## Marion Barry's Legacy

Barry's shadow hovered over the Kelly administration while the former mayor himself was never far from view (as when he was accused of having sexual relations with a prostitute in a prison visiting room within full view of other prisoners' families). As if such tawdry tales were not enough to undermine an already embattled Kelly administration, Barry returned triumphantly to Washington. The newly freed local leader managed to get himself elected in November 1992 to the D.C. Council from Washington's poorest neighborhoods in isolated Ward Eight "across the [Anacostia] river."[37]

A self-described "situationist," no human being understood the pathologies of late-twentieth-century Washington more perceptively than Marion Barry Jr.; and no politician knew better which among his many faces to turn to the electorate at any given moment. Gaining his first electoral position on the D.C. School Board as a street-smart activist, Barry managed to convince *Washington Post* editors and wealthy Ward Three white liberals that he could save the city for them when he first ran for mayor. By the end of his second term, he had turned into the darling of local developers; and he later played to racial fears among the city's African-American poor residents during the travails of his third term in office.

Barry now drew political sustenance from deeply held beliefs in Christian redemption among African-Americans to buoy his 1994 run for the District Building. His last race for mayor took place at a time when the city was arguably more polarized racially than it had been when young civil rights activists first took the helm of local administration a generation before.[38]

Many African-Americans—both rich and poor—feared a white conspiracy to take the city back from them ("the Plan," as it was known on city streets). Black residents paid special notice of an editorial in the *Washington Post* appearing on the eve of the 1994 Democratic Party primaries. The *Post*'s editors—perhaps the single most important group in bringing about Barry's initial election as mayor back in 1978—urged city voters not to be duped one more time by the city's great trickster. The paper's haughtily judgmental tone convinced many poor and middle-class African-Americans to head to the polls and support a member of their community who was once again coming under attack.

A *Washington Times* columnist, Jonetta Rose Barras, best explained the African-American community's outpouring of support for the wounded Barry in her masterful retelling of Barry's return to power, *The Last of the Black Emperors*. In 1998, Barras looked back on the fall of 1994 to observe:

> In 1994, blacks in the District chose to forget this larger history, not to mention their own dismal past with Barry. They sought a miracle to rescue them from the ineptitude of Mayor Sharon Pratt Kelly. They deeply feared the return of white dominance, and the continued physical and social decay of their city. Feigning amnesia, they reached for a known commodity. . . . With Barry in charge, most blacks believed whites couldn't force or finesse their way into the upper echelon of District politics. The mayor would protect them from the subjugation of white rule.[39]

Barry's victory brought him back to the Mayor's Office for a fourth term. In regaining the mayoralty once more, he secured the personal redemption that he so eagerly sought. But he never recovered his old mastery of the city. The world of D.C. politics had changed. His election sealed congressional disapproval of the city, eventuating in the appointment of the federal Control Board to take over management of the city.[40] Radical Conservative Republicans had gained control of the House of Representatives since Barry had left office, turning Congress into an ever more hostile force in the city's life. Competent administrators were fleeing the chaos of the city's dysfunctional bureaucracy. Some 50,000 African-Americans decamped from the city during the 1990s, largely for neighboring Prince George's County, Maryland, thereby further eroding Barry's long-term electoral base.[41] Latino and other transnational migrants—groups with little appreciation of the mayor's leadership role in the civil rights movement of the 1960s—were moving into the city in unprecedented numbers. The city was farther from financial solvency than ever.

Barry himself was worn out. New, tough-minded politicians—often mindful of the city's growing Latino population and nearly always white themselves—were beginning their own journeys to political prominence. The civil rights municipal regime, which had taken shape with such grand hopes during the early years of the 1970s, lay in ruins.

## Enter the Control Board

President Bill Clinton and congressional leaders acted quickly to put into place the requisite appointments and bureaucratic mechanisms necessary for the D.C. Financial Responsibility and Management Assistance Authority, established by Public Law 108-4 in early 1995, to begin operations. The authority's Board of Directors, which would become known in local parlance more simply as the "Control Board," assumed command over D.C. finances during the summer of 1995.

The board's five members—Chair Andrew Brimmer, Vice Chair Stephen Harlan, Joyce Ladner, Constance Berry Newman, and Edward Singletary—could not have been more different from the City Council's civil rights–era activists. Brimmer, a Harvard-trained economist who had grown up among Louisiana cotton farmers after World War II, served as a member of the Federal Reserve Board for fourteen years as well as on several major corporate boards. Harlan, the Board's sole white member and a Republican, owned a major local real estate firm. Harlan's strident criticisms of the manner in which the D.C. government functioned lent credence for some to the rumored coming white takeover of the city known as "the Plan." Ladner, a sociologist with strong civil rights credentials, had served recently as Howard University's interim president. Newman, an undersecretary of the Smithsonian Institution, and Singletary, a former vice president of Bell-Atlantic Corporation, brought both managerial credentials and institutional credibility to the group.[42] John Hill, who had overseen D.C. finances at the U.S.

General Accounting Office, was named as the board's executive director. Alice Rivlin—a former vice chair of the U.S. Federal Reserve Board, head of the White House Office of Management and Budget, founding director of the Congressional Budget Office, and senior fellow of the Brookings Institution—would preside over the board's waning years after Brimmer's departure as chair.

The board's performance has been widely regarded as a success. The District of Columbia met the financial benchmarks established by Public Law 108-4 earlier than anticipated, so that the board could pass out of existence ahead of schedule on September 30, 2001 (an event largely ignored in the wake of the terrorist attacks earlier that month on the Pentagon across the Potomac from the city, and the World Trade Center in New York).[43]

These macro-level achievements rested on a decidedly more mixed micro-level pattern of partial reform. The Control Board's eagerly anticipated efforts to bring renewed order to the D.C. public school system under the direction of a retired army general, Julius W. Becton Jr., ended in ignominious folly. A court order closed the city's public schools for nearly a month before they were deemed to be sufficiently safe to open for the 1997–98 school year.[44] The board's agents had failed to advanced their cause when one of the contractors whom they had hired to carry out school repairs parked a propane gas truck outside the windows of a classroom in which the daughter of the presiding judge was scheduled to have attended class.

The D.C. Council eventually would have to modify, suspend, or terminate several board-promoted initiatives—such as an especially inept "master license" program for consultants and freelance contractors.[45] Much of the vaunted reduction on the D.C. bureaucracy of the era was the result as much of efforts by the fourth Barry administration and D.C. Council as by Control Board staff and the city's new chief financial officer, Anthony Williams. Because it was often aloof and secretive, the Control Board effectively undermined any notion of citizen involvement in local government.

Despite miscues, the board succeeded in creating the preconditions for a dramatic shift in the local political regime from one focused on issues of civil rights, race, and social equity to an era dominated by the politics of pragmatism. In the process, Marion Barry increasingly found himself less able to hold sway over the city he had dominated for so long.

Barry still had a card or two hidden up his sleeve as he returned to the Mayor's Office for his final term. The mayor feigned conciliation with the city's new financial overseers, acting on the advice of Control Board member Ladner and an emerging Washington power broker, the D.C. delegate to Congress, Eleanor Holmes Norton.[46] Barry even formed a temporary union with Republican House speaker Newt Gingrich, in what made for one of the more bizarre alliances in American politics at the time.

Barry managed to convince the exceptionally able Michael Rogers to run the D.C. government as city administrator. Rogers, who had held senior positions in the U.S. Department of Commerce as well as in the David Dinkins mayoral administration in New York City, made impressive progress at reining in the city's budget deficit and at managing relations with Congress and the Control Board. By the end of their first year in office, Barry and Rogers had reduced the city's deficit by $150 million and trimmed its payroll by more than a thousand positions beyond what was mandated by the Control Board. Rogers successfully privatized divisions within the Department of Corrections, and he created a new Water and Sewer Authority, a separate Department of Health, and a new training program for D.C. government managers at George Washington University (the Center for Excellence in Municipal Management).

At the level of daily management, Barry and Rogers were proving to be less of an obstacle and more of a partner with the Control Board than the increasingly intransigent City Council under the mercurial Dave Clarke. Barry's public persona, however, was quite different.

Marion Barry's fourth term resembles the final battle of a prize fighter who has stepped into the ring once too often. A master of race politics and the rhetoric of defiance, Barry continued to blame others for the city's problems. This time around, he no longer had the energy, wisdom, discipline, and staying power to gain the upper hand. His dalliance with Gingrich predictably ended with few results for either Barry or the city. Increasingly jealous of Rogers's successes and growing reputation, Barry eventually brushed his city administrator aside by late 1997.

Tactics that had given Barry an edge in the past now failed to provide advantage. Bigger game had entered the scene, not the least of which was a Republican-dominated Congress, which extended Brimmer and the Control Board ever more power and authority whenever Barry seemed to be gaining an advantage. There was also a fresh upstart on the block: Anthony Williams.

## Mr. Bow Tie

Anthony Williams is an enigma. The son of a poor family from Los Angeles, he grew up with seven brothers and sisters before making his way to

Yale College. Awkward and uncomfortable with people, the nerdish Williams eventually graduated magna cum laude in political science from Yale before continuing on to earn his law degree from Harvard, along with a master's degree in public policy from Harvard's John F. Kennedy School of Government. Though a self-proclaimed conscientious objector, he became a pilot with the 354th Tactical Fighter Wing of the U.S. Air Force. And though he was painfully soft-spoken, he managed to get himself elected to the city council of New Haven, Connecticut.

A man of unquestioned professional expertise and personal integrity, Williams nonetheless has proven himself capable of incomprehensible ethical gaffs, as when his 2002 reelection campaign was marred by charges of corruption surrounding improperly filed petitions to have his name placed on the ballot. Williams's hired hands at the time submitted petitions with more than 4,000 disqualified signatures (including forged signatures from such prominent nonresidents as Arnold Schwarzenegger and British prime minister Tony Blair, as well as various and sundry pop music and film stars). The normally timid D.C. Board of Elections threw him off the ballot, so he was forced to win reelection through a write-in campaign.[47]

Williams behavior became even more inscrutable once District of Columbia voters returned him for a second term as mayor. He began to travel extensively, causing some to speculate that he was out of the city more than he was present. He disengaged from public discourse, remaining eerily aloof in his meetings with constituents and unengaged in sessions with other politicians. His one major achievement—convincing Major League Baseball to move the Montreal Expos to Washington—almost unraveled because, as will be discussed below, he initially failed to enter into the backroom wheeling and dealing required to move the requisite public financing bills through the D.C. Council.

The many contradictory facets of Williams's personality rest uncomfortably within a tall, lanky frame hidden behind a trademark bow tie. To local voters—and, perhaps yet again most important, the Editorial Board of the *Washington Post*—"Mr. Bow Tie" was the very antithesis of Mayor Barry. Barry must have thought so as well when he lifted Williams from his post as the chief financial officer of the U.S. Department of Agriculture to appoint him as the city's chief financial officer in October 1995.

For perhaps the first time in his long career, Marion Barry had misjudged a potential opponent. Whether it was Williams's geekish style, his plodding demeanor, his Ivy League bow ties, or more simply the aging Barry's weakening political touch, the mayor allowed Williams to outflank him at critical moments. For example, the upstart artfully grabbed praise and recognition for allegedly leading the city back from the brink of bankruptcy. Williams also established his own special relationship with two key players on the local scene: North Carolina Republican representative Charles Taylor, who served as chair of the House subcommittee overseeing D.C. appropriations; and the editors of the *Washington Post.*

By 1996 and 1997, Washingtonians increasingly awoke to find themselves greeted by media reports of Williams's financial wizardry. D.C. congressional delegate Eleanor Holmes Norton, Control Board Chair Andrew Brimmer, and Mayor Barry all suggested that the chief financial officer was hardly the magician that he seemed to be to so many.[48] Nevertheless, it would be Barry who was left to pack up for retirement by early 1998. Williams, for his part, became the newly anointed mayor of the District of Columbia in the primaries and elections later that year.[49]

The 1998 municipal elections marked a moment of regime change for Washington. Certainly, Mr. Bow Tie cut a very different image from his Dashiki-attired predecessor. Changes far more substantial than choice of clothing were under way as well. By the time the electoral dust had settled, the civil rights municipal regime of Barry, Clarke, Tucker, Wilson, Hobson, and so many others no longer existed. A new pragmatism ruled city affairs, presided over by a green-eyeshaded accountant-mayor together with an even more remarkable and less expected gaggle of arrivistes in the District Building: a white-majority City Council.

If African-Americans had voted in 1994 for Marion Barry Jr. to prevent whites from seizing control of local government, Barry failed his supporters four years later. Barry launched another election campaign in 2002 for an at-large seat on the D.C. Council. That effort swiftly collapsed after police found the former mayor parked in the middle of the night under a bridge over the Anacostia River with an unidentified woman other than his wife.[50]

In 2004, Barry attempted yet again to get onto the D.C. Council by running against incumbent Ward Eight council member Sandy Allen, who happened to have been one of Barry's former campaign managers.[51] The local media loved Barry's latest run, playing up his statements that he was being forced back into public life after becoming "horrified about what's happening."[52] Initially, times appeared to have changed, as Barry proved himself to be a surprisingly inept campaigner.[53]

Yavocka Young, a member of an Advisory Neighborhood Commission (ANC) in Barry's own Ward Eight, expressed the considerable displeasure of many in D.C. politics with "Mayor-for-Life" Barry's presumptive claim of a seat on the D.C. Council. "He tends to nurture those people who are down on their luck," noted Young, who had purchased a house in Ward Eight's Anacostia area in 1991 just as Barry was in the painful throes of a personal collapse.[54] Young continued, "But we can't just soak up all the social services the city has to offer. We have to be able to get in the game."

The canny Barry knew better than his critics that the city's dispossessed had not bought into the new pragmatism at the District Building. Ward Eight voters—watching city services wane "east of the [Anacostia] river," feeling the pressures of incipient gentrification, ignored by their mayor, resentful of the visible abandonment of a civil rights–era city agenda, sensing that "the Plan" of whites to seize control of the city away from blacks was playing itself out—rallied behind their long-term standard-bearer and elected Barry to be their representative on the D.C. Council. As DeWitt Kinlow, a former president of the Ward Eight Democratic Club, explained to the *Washington Post,* "Voting for Marion was a statement from us to them. We put a person in who remembers us. We put someone in who would tell the Man the way it is. People in Ward 8 don't often get to make that kind of statement. You put Marion back in office, and we will be heard."[55]

Barry won a landslide victory, proving yet again his special tie to the city's poor and powerless. He carried 61 percent of the vote (to Allen's 23 percent) in Washington's poorest ward during the all-important September 2004 Democratic Party Primary. The irrepressible "mayor-for-life" had tapped a deep seething tide of anger that lay just below the surface pragmatism of the city's new leaders.[56]

The city's political culture had changed profoundly during the years since Barry had last left office. Cable television coverage of council deliberations highlighted a new practicality that had been absent from the District Building during the first quarter-century of home rule. On the same day that Marion Barry won the Democratic Party Primary for the Ward Eight council seat (guaranteeing his election in the predominantly Democratic city), a local business executive and civic activist, Kwame Rashaan Brown, defeated longtime D.C. Council member Harold Brazil for an at-large seat. The pragmatic community activist Vincent C. Gray simultaneously emerged victorious in Ward Seven over the incumbent Kevin Chavous.[57]

Brown, who had held a number of jobs in business as well as the U.S. Department of Commerce, is a product of D.C. public schools who graduated from Morgan State College and earned a master's in business administration from Dartmouth College's Amos Tuck School of Business.[58] Gray, the chair of the Ward Seven Democratic Committee, similarly went to local schools and then graduated from George Washington University.[59] He served as the head of the D.C. Department of Human Services and won acclaim as the executive director of Covenant House, a social service provider and advocacy organization.

Brown and Gray had been active in neighborhood organizations. Both had distinguished professional careers and educational accomplishments; and both were deeply rooted in their communities. Brown and Gray fit the profile of the emerging D.C. regime of municipal pragmatism. They both ran—as did Barry for that matter—on platforms of trying to make the D.C. government more effective.

## Advisory Neighborhood Commissions and the Politics of Municipal Pragmatism

The story of changes on the D.C. Council is as remarkable as those of the demise of Barry and the rise of Williams as the D.C. mayor. Throughout the 1990s, weary city residents had been losing heart, hope, and interest in a political process that increasingly pushed voters aside in favor of seemingly uncontrollable events and unaccountable decisionmakers. Leading voices for the old race-based politics of the previous quarter-century were disappearing one by one, replaced by those intent on balancing budgets and streamlining municipal administration. New politicians saw a moment to gain prominence; and seized that opportunity to reshape the city's electoral scene.

Barry, Clarke, and other successful D.C. politicians had leaned heavily on African-American church leaders to secure the votes necessary for victory. The election of a white homosexual Republican to a citywide post would have appeared to have been the least plausible outcome of any election right up until the moment of Dave Clarke's funeral. But that unlikely outcome had become a reality before the year was over.

Home rule established a system of thirty-seven Advisory Neighborhood Commissions to review policies and to consult with city officials on a range of local issues including traffic, parking, recreation, liquor licenses, economic development, police protection, sanitation, and other municipal services. The ANCs have always been something of a poor stepchild in D.C. governance, having minimal budgets and no direct power. Many of the unremunerated ANC seats remained vacant or uncontested. A few ANCs—especially in some of the more wealthy wards in the city's Northwest quadrant—developed into vehicles for local debate. Community activists grabbed whatever minimal power and authority might be vested in their

ANCs to fight off unwarranted interference from "downtown" and to demand improved services for their neighborhoods.

With little turnover in D.C. Council positions—and with no possibility of defeating Mayor Barry in sight—observers of D.C politics tended to ignore the ANCs. The commissions nonetheless produced a generation of leaders much more concerned with pragmatic problem solving at the neighborhood level. If the first home rule elite had emerged from the early School Board elections of the late 1960s and early 1970s, members of Washington's post-1998 elected municipal elite cut their teeth on the ANCs.

Ward Two council member Jack Evans initially showed this path to city office when he was elected to the council to replace John A. Wilson, who had just won the race for council chair in 1990. Future council ward members Jim Graham, Adrian Fenty, and Sandy Allen, as well as at-large members Phil Mendelson and David Catania—like Evans—began their political careers on their neighborhood's ANCs. Emerging from the most local of local political institutions, this new breed of council member has proven to be more concerned with the nitty-gritty politics of trash collection, public safety, and zoning than with the larger philosophical issues of social and racial justice that so drove earlier D.C. politicians such as Barry, Clarke, Hobson, and Tucker or the public finance that so captivates Williams.[60]

D.C. Council politics have been shaped by another peculiarity of the home rule design. No single political party can nominate a sufficient number of candidates to win all the council's four at-large seats. In other words, Congress ensured that the D.C. Council would have members of parties other than the Democrats. For many years, these seats had been held by Independents and members of the D.C. Statehood Party (first Hobson, and later Hilda H. M. Mason). With Clarke gone, a shift of one at-large member would free up a seat for an off-year election term that would expire in 1998. A Republican theoretically could win in 1997, subsequently slipping into one of the two seats reserved for a non–majority party member (then occupied by aging Statehood Party council member Mason). David Catania, an aggressive neighborhood activist from the Sheridan-Kalorama ANC, saw his opportunity and made the run.

The onetime schoolteacher and former School Board president Linda W. Cropp initially ran as a Democrat for one of the council's at-large seats in 1990. She emerged as the natural successor to Clarke, being named as acting chair immediately upon his death. D.C. voters confirmed this decision in a special election held in August 1997. Cropp's own at-large seat thus came up for grabs, to be filled by a special election on the first Tuesday of December.

The campaign for Cropp's seat on the council was an especially dispiriting affair. The city was deeply mired in the increasingly degrading spectacle of Barry's last grasp for respectability and power. The Control Board was withdrawing into itself following the demeaning spectacle of a school year that had begun weeks late as a result of court orders declaring far too many classrooms unsafe. The District of Columbia appeared to be headed for financial and political oblivion.

The Democrats, for their part, struggled to find a candidate to run for Cropp's now-vacated at-large seat. They eventually put forward the candidacy of former D.C. Council chair Arrington Dixon, who had last held local office in 1982, when his reelection bid for council chair failed before the candidacy of a rising Dave Clarke. Dixon's most recent claim to municipal fame seemed to be that he had once been married to luckless Mayor Sharon Pratt Dixon Kelly.

Sensing a race that was a foregone conclusion, and preoccupied with a fast-approaching holiday season, 93 percent of the city's registered voters stayed away from the polls on election day. Catania, a neighborhood and gay rights activist, campaigned hard, successfully mobilizing his core base of supporters. When all the votes were counted, Catania had collected 10,818 votes citywide to Dixon's 9,621. Dixon lost despite impressive majorities in four of the city's eight wards.[61] The unthinkable had happened—a white Republican lawyer, and a homosexual at that—had defeated a mainline Democrat in a citywide race. With the guarantee of the Republican's nomination for one of the two "minority party" at-large seats, Catania appeared likely to become a long-term fixture on the council (as, indeed, he has).

Catania eventually broke ranks with his own party during the 2004 presidential race, angered by President George W. Bush's support of an amendment to the U.S. Constitution that would ban same-sex marriage.[62] His place as an Independent on the D.C. Council seems no less secure for having left the Republican Party in a city that voted 89.2 percent for the Democratic Party presidential candidate John Kerry, as opposed to 9.3 percent for President Bush.[63]

## The Election of 1998 and the Emergence of a Pragmatic Municipal Regime

Many leaders of the African-American community, especially the powerful ministers and bishops of local churches, remained shocked and silent at these events. Marion Barry quickly calculated that further shifts in local politics might be afoot. Sensing the impending possibility of a majority white D.C. Council for the first time in modern history, the mayor began to speak out against white candidates for local office. It was Williams, however, who set the tone for the next round of city elections.

With Barry neutralized by late springtime, Williams used money and the press to beat back a challenge from Ward Seven council member Kevin Chavous, a Howard University–trained lawyer and education advocate. Williams garnered precisely 50 percent of the vote in the September Democratic Party primaries in a seven-candidate race in which Chavous placed second with 35 percent (drawn largely from the old civil rights constituency that had previously dominated local elections). Williams eventually went on to trounce the Republicans' perennial mayoral candidate, longtime council member Carol Schwartz, in the November general election, garnering 89,573 votes to Schwartz's 41,072.[64]

In one of the many ironies of this unusual election season, the white Republican Schwartz had stronger ties to the city's previous civil rights municipal regime than did black Democrat Williams. Born in Mississippi and raised in Texas, Schwartz moved to Washington following graduation from the University of Texas to teach special education in the public schools. She served two terms on the original elected School Board with Barry and others activists before being elected to her minority party at-large council seat for the first time in 1984. She stepped away from politics following the death of her husband in 1988 but returned to the council in 1996. Much beloved throughout Washington, the affable Schwartz is widely viewed as a passionate D.C. advocate.[65] In a city in which registered Democrats outnumber Republican voters by more than ten to one, Schwartz's final tally in 1998— and again in 2002, when she increased her total to 45,407 votes to Williams's declining support of 79,841 ballots[66]—was a considerable accomplishment.

The fall's most heated and racially charged contests proved to be within the Democratic Party's primaries for the D.C. Council. Democrat Phil Mendelson won a hotly contested at-large seat on the D.C. Council with just 17 percent of the vote in the all-important September Democratic Primary election. Another five candidates, all African-Americans, polled more than 12 percent of the vote, creating the space for Mendelson to eke out victory.[67] Mendelson—a native of Cleveland who had come to Washington in 1970 to pursue a degree in political science from American University—served as a legislative aide to former Ward Three council member Jim Nathanson and then for Council Chair Clarke. He gained his first elective office in 1979, when he ran for his local ANC.

Mendelson had been involved in local affairs from a "bottom-up" perspective, having been a leader in the prominent Committee of 100 on the Federal City.[68] He personified what was becoming a new norm for the council: pragmatic, neighborhood-based community activists who moved from outside Washington to pursue their professional careers before becoming engaged in local politics. As with Catania's and Williams's victories, Mendelson's triumph would have been improbable just a few years before.

The pivotal race in bringing about regime change took place in centrally located Ward One, the city's most densely populated and ethnically diverse council district. Transnational migrant votes would tip the balance before the electoral dust would settle on a new municipal regime.

Ward One had been represented by only two people since 1974: Dave Clarke and Frank Smith Jr., who had been first elected in 1982 when Clarke moved on to take over the council's chair. Smith, a onetime leader of the Student Nonviolent Coordinating Committee, arrived in the city in 1968. He was very much part of the same local elite that had emerged with home rule, sharing a background in the civil rights movement and in SNCC with Barry, Clarke, and others.

Smith angered many of his longtime labor union supporters by working on the D.C. Council with the Control Board to straighten out the city's finances. Moreover, Ward One—more than any other in the city—was beginning to feel the city's profound demographic shifts discussed in chapter 2. Containing Mount Pleasant and Adams Morgan, the ward's longtime African-American communities were giving way to new white and Latino residents.[69] Smith had become vulnerable.

Jim Graham, a nationally recognized AIDS activist and longtime executive director of the city's Whitman-Walker Clinic (a leading HIV/AIDS treatment center), had worked with Smith in the past. Graham arrived in Washington following graduation from Michigan State University and the University of Michigan Law School to serve as a clerk for Earl Warren, chief justice of the United States. He stayed on, earning additional law degrees from Georgetown University and becoming a staff attorney with the U.S. Senate Governmental Affairs Committee under its chair, Senator Abe Ribicoff. His AIDS work brought him into local affairs. Graham effectively mobilized support of his ward's large gay community, piecing together a strong coalition with Latino and other neighborhood-oriented activists.[70] Graham collected 48.6 percent of the vote in the September Ward One Democratic Primary, to Smith's 32 percent.[71]

Williams, Catania, Mendelson, and Graham made their ways as outsiders to the traditional civil rights activist core of pre-1998 home rule politics (though, they all—even the Republican Catania—would have been considered to be strong liberals elsewhere in the United States). Encountering the city's dysfunctionality in their jobs and daily lives, this new breed of D.C. politician turned to community and neighborhood activism. Their agenda, unlike that of the Barry-Clarke generation, became oriented fundamentally toward solving problems. They wanted their city to work. With the exception of Tony Williams, they were all white.

Given the Democrats' overwhelming registration advantage, their victories in the party's nomination primaries guaranteed success in the November general election. Without some unforeseen intervention, observers quickly understood that the D.C. Council would become majority white for the first time should Catania unseat the increasingly feeble at-large Statehood Party council member Hilda H. M. Mason.

Mayor Barry was the first to express outrage. Barry announced his refusal to support any white Democratic candidate for the D.C. Council the day following the primary. "I think the council ought to reflect the makeup of the city. I am not a blind Democratic supporter," Barry told reporters. Barry continued on by attacking white council members for their efforts to cut funding to his programs. Whites and blacks are different, Barry went on, "in terms of culture, in terms of philosophy."[72] He concluded by arguing that elected officials from the same "culture and race as the dominant population make for more thoughtful policy development and implementation."[73]

Barry's comments polarized the city's political scene along racial lines for some and were cast off as inappropriate and offensive by others. Many African-American commentators, such as the always contemplative Howard Croft (then an administrator at the University of the District of Columbia), were less alarmist. "There is a real dilemma here," observed Croft. "How do you square the interests of a group to see itself fully represented with the desire to treat the candidates as real live individuals who ought to be judged by who they are and what they can do?" Croft, who had lost to a white woman, Sharon Ambrose, in a special primary for the Ward Six council seat the year before, endorsed Mendelson based on the candidate's liberal philosophy and record.[74]

The November general election completed the council's membership turnover, as Catania defeated Mason while both Graham and Mendelson enjoyed commanding victories.[75] Six of the council's thirteen members had been elected over the course of the previous two years, with a majority of council members gaining their seats since 1994 (see tables 5.1 and 5.2).[76]

*Table 5.1. Members of the District of Columbia Council, January 1, 1997*

| Post | Name (Party) | First Elected to Council | Left Council | Race |
|------|--------------|--------------------------|--------------|------|
| Chair | David A. Clarke (D) | Nov. 5, 1974 | March 27, 1997 | White |
| At-large | Hilda H. M. Mason (Statehood) | Nov. 7, 1978 | Nov. 3, 1998 | African-American |
| At-large | Linda Cropp (D) | Nov. 6, 1990 | | African-American |
| At-large | Harold Brazil (D) | Nov. 5, 1996 | Dec. 31, 2004 | African-American |
| At-large | Carol Schwartz (R) | Nov. 5, 1996 | | White |
| Ward 1 | Frank Smith (D) | Nov. 2, 1982 | Sept. 15, 1998 | African-American |
| Ward 2 | Jack Evans (D) | April 30, 1991 | | White |
| Ward 3 | Kathleen Patterson (D) | Nov. 8, 1994 | | White |
| Ward 4 | Charlene Drew Jarvis (D) | July 1, 1979 | Sept. 12, 2000 | African-American |
| Ward 5 | Harry Thomas Sr.(D) | Nov. 4, 1986 | Sept. 15, 1998 | African-American |
| Ward 6 | Vacant | — | — | — |
| Ward 7 | Kevin P. Chavous (D) | Nov. 3, 1992 | Dec. 31, 2004 | African-American |
| Ward 8 | Sandy Allen (D) | Nov. 5, 1996 | Dec. 31, 2004 | African-American |

*Table 5.2. Members of the District of Columbia Council, January 1, 1999*

| Post | Name (Party) | First Elected to Council | Left Council | Race |
|------|--------------|--------------------------|--------------|------|
| Chair | Linda Cropp (D) | Nov. 6, 1990 | | African-American |
| At-large | David A. Catania (R) | Dec. 2, 1997 | | White |
| At-large | Phil Mendelson (D) | Nov. 3, 1998 | | White |
| At-large | Harold Brazil (D) | Nov. 5, 1996 | Dec. 31, 2004 | African-American |
| At-large | Carol Schwartz (R) | Nov. 5, 1996 | | White |
| Ward 1 | Jim Graham (D) | Nov. 3, 1998 | | White |
| Ward 2 | Jack Evans (D) | April 30, 1991 | | White |
| Ward 3 | Kathleen Patterson (D) | Nov. 8, 1994 | | White |
| Ward 4 | Charlene Drew Jarvis (D) | July 1, 1979 | Sept. 12, 2000 | African-American |
| Ward 5 | Vincent Orange (D) | Nov. 3, 1998 | | African-American |
| Ward 6 | Sharon Ambrose (D) | April 29, 1997 | | White |
| Ward 7 | Kevin P. Chavous (D) | Nov. 3, 1992 | Dec. 31, 2004 | African-American |
| Ward 8 | Sandy Allen (D) | Nov. 5, 1996 | Dec. 31, 2004 | African-American |

Together with Tony Williams as mayor, the D.C. Council would bring about a new municipal regime in the District Building. Local politics were increasingly dominated by a pragmatic concern over city services and neighborhood quality of life as well as the exigencies of city planning rather than larger issues of social and racial justice.[77]

## Take Me Out to the Ball Game—or Not?

Few Washington residents, let alone area suburbanites, paid much attention to the profound changes that were taking place at the District Building. There seemed little reason to do so as the issues of concern on the D.C. Council became more narrowly focused on how the city would function. All too often, the council could do little more than rail away from the sidelines at the major diminution in services provided by authorities and agencies beyond its control (e.g., the failure of the D.C. Water and Sewer Authority to inform city residents of dangerously high levels of lead in local drinking water, the closure of major streets by competing federal police forces in response to alleged terrorist threats without consultation with D.C. officials, the marked deterioration in subway and bus service mandated by the Washington Metropolitan Area Transit Authority).[78] Mayor Anthony Williams found more and more reason to be away "promoting" D.C. interests, or speaking to this or that out-of-town group about municipal finance.[79] Most observers, therefore, were caught off guard when an issue of major city and regional interest—the construction of a major league ballpark to be built solely with D.C. public funds—hinged on council deliberations.[80]

The issue of Major League Baseball's return to the Washington region had been an emotional one ever since the Washington Senators had left town for Texas in 1971. Washington had a long, though largely undistinguished, place in the history of America's national pastime dating back to the mid-nineteenth century. The original Senators, who played in various incarnations in the National League at the end of the nineteenth century, had been one of the founding teams of the American League in 1901.[81] Though they won the World Series only once (in 1924) and the American League Championships but three times (1924, 1925, 1933), the Senators nonetheless fielded some fine teams in the early twentieth century behind Walter "Big Train" Johnson, one of the best pitchers in the history of the game. The Homestead Grays, perhaps the finest team ever to take the field in the Negro Leagues of the 1930s, and 1940s, with such legendary stars as the powerful slugger Josh Gibson and the fleet-footed Cool Papa Bell, played some of their best ball at Washington's Griffith Stadium despite a long-standing connection to Pittsburgh.[82]

The postwar period was a difficult one for the Senators. Their owner, Calvin Griffith, was among the last to sign African-American ballplayers.[83] In 1961, he moved his franchise to Minneapolis just as the team had begun to regain preeminence. Renamed the Twins, the team would play in the 1965 World Series with many of the same members who had been on the roster in Washington.[84] Griffith had long complained about Washington's changing racial composition, explaining his team's relocation by suggesting that he could not successfully run a baseball club in a predominately black city.[85] Major League Baseball responded by placing a new "expansion" club in Washington, a largely hapless crew that would abandon Washington after the 1971 season with only one winning year to its credit to become the Texas Rangers.[86] For many living in the Washington region, the next three decades would become a painful series of unsuccessful attempts to convince big league baseball to return.

Meanwhile, the Montreal Expos, a National League team created in 1969 and named after Mayor Jean Drapeau's 1967 World's Fair, had entered its own cycle of decline.[87] The team's best season—1994—was interrupted by a players' strike, bringing the year to a close without postseason play at a time when the Expos had the best record in the major leagues.[88] Embittered Montreal fans began to desert the team, initiating a vicious cycle of falling revenues and a selling off of talented players to recoup the losses, which eventuated even greater fan flight.

A series of disastrous ownership moves as well as the failure of the City of Montreal and the Province of Quebec to build the team a new stadium landed the Expos in receivership just as Major League Baseball was looking to contract by shedding "small-market" teams such as Montreal and Minnesota.[89] Court action and player union negotiations blocked the owners' efforts to close franchises. Consequently, in February 2002, the twenty-nine remaining teams collectively purchased the Expos and put them on the bidding block.[90]

The Expos began to play some of their "home" games in San Juan and improved their onfield performance during 2002. Success was followed by further collapse in 2003.[91] Washington mayor Williams did his best to position his city to win a bid for the team, eventually overcoming competition from San Juan; Northern Virginia; Portland, Oregon; Las Vegas; and Monterey, Mexico—among many suitors.

Williams and representatives of Major League Baseball announced an agreement to move the Expos to Washington on September 29, 2004, just as the regular season was drawing to a close. Major League owners approved the move on December 3, just as the issue of a new stadium was before the D.C. Council.[92]

Williams had committed the city to building the team a new stadium during his negotiations with the owners. The economics of public sports facilities in the United States are hotly contested. A strong case has been made for the beneficial impact of privately built arenas in downtown locations— such as the MCI Center, which had opened in 1997 at Seventh and F Streets, Northwest. The entire neighborhood around the MCI Center has been revitalized, becoming one of the city's major entertainment districts and home to an ever-increasing number of residents. Abe Pollin, a local developer and owner of the National Basketball Association's Wizards professional team as well as a major investor in the National Hockey League Capitals team, built the $260 million arena with his own funds together with private loans. The city topped off Pollin's investment with $60 million worth of improvements in the surrounding public infrastructure.[93]

Baseball and football stadiums are more expensive, used less frequently, and generally considered to be a more risky public investment. For example, the Washington Redskins football team moved to suburban Largo, Maryland, in 1997 after the D.C. government had failed to provide what its owners considered to be a suitable site and public financing. Their new stadium, FedEx Field, is the largest in the National Football League and cost $250.5 million ($180 million of which was provided by private financiers).[94]

Just as the D.C. government entered into negotiations with Major League Baseball, city officials were simultaneously working with a private developer to build a new stadium for the city's champion professional soccer team, D.C. United, directly across the Anacostia River from the proposed ballpark. The cost of this considerably more modest stadium would be borne by the private sector. The soccer complex would be embedded in the surrounding community, with training, retail, recreational, and concert facilities available to city residents (including the city's burgeoning transnational migrant community).[95]

Overall, the record on baseball stadiums as neighborhood revitalization magnets is generally mixed. At the height of the D.C. Council debates, the *Washington Post*'s Peter Whoriskey visited two recent examples of efforts to use ballparks to reinvigorate downtown neighborhoods, Safeco Field in Seattle and Coors Field in Denver. Whoriskey concluded that the experiences of Seattle and Denver show just how difficult it is to predict a stadium's role as a catalyst for neighborhood revitalization. The ballpark projects in Seattle and Denver are roughly similar to the one that Washington's leaders are considering. Both stadiums relied largely in public funding, both were built in run-down areas near downtown and both inspired hopes that they would spur the turnaround of those neighborhoods. Now, the Denver and Seattle ballparks symbolize the two extremes in a national debate over the economic benefits of building stadiums. Boosters of stadium projects cite Denver and a few other cities as success stories, while skeptics list a group of cities that includes Seattle.[96]

At best, supporters of such projects argue that the combination of psychic and financial benefits to be derived from being a "big league town" offset the short-term financial costs of securing teams through public financing of sports venues.[97] The deal cut between Mayor Williams and Major League Baseball was far from such optimistic projections.

Among the conditions set down by the baseball owners to bring the Expos to Washington was a commitment by the D.C. government to construct a new stadium for the team by March 2008 at public expense. The stadium—which would become a focal point in a broad waterfront redevelopment project along the Anacostia River—would cost the city an estimated $584 million.[98] Lucrative cable television rights for the team would be held by Baltimore Orioles owner Peter Angelos.[99] Moreover, the city would pay the team, now renamed the Washington Nationals, $19 million each year after March 2008 that the stadium had not been completed.

Within weeks, the city's chief financial officer, Natwar M. Gandhi, estimated that the cost projections for the project had been undervalued by $100 million.[100] The euphoria so evident at the September 29, 2004, announcement of the Expos' relocation to Washington began to wear thin with each new revelation about the cost to city taxpayers.

Williams sought to take the sting out of the financial package for many Washington residents by proposing that the cost be borne by a special citywide levy on businesses. As the costs of the proposed stadium—and hence, the size of the proposed taxes encumbrance on businesses—began to grow, representatives of some leading Washington businesses sought out Council Chair Linda Cropp.

On November 5, 2004—just days after stadium opponents Marion Barry, Kwame Rashaan Brown, and Vincent C. Gray had won their seats on the council—Cropp held a news conference calling for a less expensive facility to be built with private financing near the existing Robert F. Kennedy Memorial Stadium.[101] Baseball owners immediately rejected the plan.

The first of two scheduled votes by the council were postponed until November 30, as D.C. leaders feverishly sought to reconfigure the financial plan more in the city's favor. Cropp joined two colleagues in abstaining during this early vote, explaining that neither could she support Williams's plan nor did she wish to foreclose on the opportunity for the Expos to move to Washington. The proposed plan limped forward after six council members approved the first reading of the required legislation that evening, four voted against, and three (including Chair Cropp, abstained).

Williams finally began to intervene the following day to renegotiate the deal between Major League Baseball and the city government.[102] Baseball proponents faced a hard deadline at the end of December for the second vote. Not only did the initial agreement with the baseball owners expire on December 31, 2004, but three baseball supporters (Allen, Brazil, and Chavous) would be replaced by three opponents (Barry, Brown, and Gray) the next day.

The second, seemingly decisive vote was scheduled for December 14, 2004. Supporters, such as Jack Evans, believed that a workable deal had been struck.[103] Some readers of the *Washington Post* who were in favor of the stadium deal may have been chastened when they opened their papers that morning. On the opinion page opposite the editorials, Marion Barry wrote, "As much as we all want baseball games to take our kids to—and I do—we should not allow our enthusiasm to sanction mendacity, and, in particular, to give the green light to powerful new corporate interests to treat the city and its people like second-class supplicants and pawns."[104]

The marathon D.C. Council session lasted more than a dozen hours. Late in the proceedings, after 10:00 p.m. that night, the deal fell apart seemingly inexplicably as Cropp announced her inability to support the proposed legislation. The council eventually passed legislation that would require half the stadium's financing to come from private sources. Major League Baseball closed the Nationals franchise the next day.[105]

The prospect of a total failure prompted much hand-ringing and intense behind-the-scenes negotiations among all parties. Just a week later, a final bill requiring that the mayor and baseball owners seek (but not necessarily obtain) up to half the financing for the proposed stadium from the private sector passed by a vote of 7 to 6, with Cropp shifting sides once again to join with Evans and other baseball supporters.[106] The council removed various penalties for failure to finish the stadium's construction by opening day

2008, and it added provisions specifying that local businesses be used to build the ballpark. "Beaming like a kid," Mayor Williams signed the final bill into law on December 29, three months to the day after initially having announced baseball's intention to move the Montreal Expos to Washington.[107] The team took the field weeks later.[108]

The D.C. Council's behavior throughout this saga—and Cropp's performance in particular—enraged many observers. Several media figures found the council and Cropp to be overly vexatious. Thomas Boswell, an award-winning baseball columnist for the *Washington Post*, harumphed mightily that "the game wonders if it has jumped in bed with a nut case."[109] Boswell, who evidently has not spent much time considering the often unseemly nature of legislative democracy, mused further:

> Part of the attention has come in the form of laughter and even mockery that Williams and his council can't stay on the same page or even in the same book on a huge deal. In major negotiations, who speaks for Washington? For now, the answer is, "You never know."[110]

Marc Fisher, a *Washington Post* columnist who covers metropolitan issues, wrote for many who were increasingly dismissive of the D.C. Council's performance. Fisher frequently danced along the line of appropriateness, describing Cropp's actions as a "curious crusade," darkly alluding to the return of Marion Barry to the D.C. Council, while scoffing at Cropp's actions as being "all about Linda."[111] Such viciously derisive contempt for the city of Washington all too frequently moved beyond disagreement over the serious issues of public policy and finance posed by the stadium proposal. A baseball fan named Jeff went so far as to set up an "I Hate Linda Cropp!" Web site; while another stadium proponent launched a Web log titled "Linda Cropp Sux."[112] At its worst, such commentary reminded some long-term city residents of the racial bating that so often marred the city's history.[113]

By contrast, another *Washington Post* sports columnist, Sally Jenkins, undoubtedly expressed the views of many opposed the stadium deal when she wrote that she was "aghast" at what she saw:

> I saw a bunch of baseball honchos, led by [baseball commissioner Bud] Selig and his dealmaker [Chicago White Sox owner] Jerry Reinsdorf, who treated the nation's capital like a sucker ripe for rooking, and took its mayor for a sap. I saw a mayor who made promises to both sides that

he couldn't keep and agreed to terms he never should have—and that Rudi Guliani surely would never have tolerated.

I saw a D.C. Council that knows its mayor all too well and sniffed warily at the reek of this deal, even as some of its members stood on the podium next to him and put Nationals caps on their heads. Several council members who were at the celebration, including Cropp, said in no uncertain terms they needed to study the stadium financing plan before they decided how to vote on it. And when Cropp did look at it, and decided she didn't like it, she was called a backstabber.[114]

Baseball proponents and opponents alike detected a defining moment in the city's political life, the point in time when the D.C. Council emerged as a significant element of the city's life, and when Mayor Williams's flaws appeared fatal. More profoundly, the effort to reinvent the Montreal Expos as the Washington Nationals did not redefine D.C. politics so much as it brought into broader view changes that had been taking place at the District Building since 1998.

For the first time in memory, a major public policy discussion in D.C. politics transcended racial boundaries (at least in so far as city politicians were concerned, though the scurrilous comments of many largely suburban baseball fans crossed that line at times). Four of the council's six steadfast stadium supporters were African-Americans (not counting Chair Cropp, who switched sides throughout the debate); the project's most visible supporter—Jack Evans—was white. Five of the six consistent opponents on the 2004 council were white; all three newly elected stadium opponents who would take office in January 2005 were African-American.

Just as important, according to opinion polls, more than two-thirds of D.C. voters representing every neighborhood as well as all ethnic and racial groups within the city opposed using only public funds to build a ballpark by the time the debate had run its course.[115] The council members knew their constituents better than media sports commentators did. They also knew the local business community. Various groups submitted bids to privately finance all or part of the stadium project, often in return for development rights in the immediate surrounding area. None of these plans proved viable. Cropp was philosophical over the result, telling the *Washington Post,* "You've got to try for private financing because if you don't, you're definitely not going to get any. All in all, I'm thrilled in hindsight with the position I had taken."[116]

A nonracial division within the D.C. Council and the city's body politic appears to have been at work on the council and throughout the city more generally. Four of the six stadium proponents on the council were elected before 1997, as opposed to only two of the six opponents (to which one can add the three incoming council members, who are on record as being opposed to the baseball financing package). Washington's great baseball debate of 2004 marks the moment when the shift from a civil rights municipal regime to a pragmatic municipal regime becomes most demonstrable.

Transnational migrants and their representatives were visibly absent from the fracas surrounding baseball. Their spirit, however, was ever present. Arriving in the city in large numbers just at the moment of municipal regime shift, transnational migrants helped mold a broader context within which the middle ground of pragmatism could emerged from within the racial politics of the past.

The D.C. Council had spent months debating less visible but no less important issues before September 29, 2004. The normal flow of previous council business shaped the dynamics of the divisions and conflicts that burst into view as the stadium debate gained full fury. Many of those previous concerns were directly related to the growing presence of transnational migrant communities in the city. In this very real way, these migrants helped to mold the a new pragmatic municipal regime that became consolidated with the legislative battles throughout the fall of 2004.

## A Political Agenda No Longer Written in Black and White

With the exception of the 1998 Ward One council race, in which Latinos actively supported Jim Graham, Washington's transnational migrant community played only a background role in bringing about this profound shift in Washington's political life. Graham was alone among of the key players in the city's politics to have been born outside the United States (in Wishaw, Scotland). Few city leaders had more than cursory contact with Latino and transnational migrant concerns. The arrival of politicians elected on platforms favoring practical municipal concerns nonetheless created political space for transnational migrants that had not existed previously.

The absence of attention to the city's foreign-born population had never been more striking than in the months and years following the 1991 Mount Pleasant riots. Those disturbances might have been expected to elevate general awareness about the city's growing Latino community. Yet a few commissions and halfhearted, underfunded gestures aside, remarkably little changed. Latino concerns suffered from inattention and ineptitude under Mayor Kelly's administration.

The return of Marion Barry had refocused local politics on the primal issue of race. The overwhelming budgetary exigencies of the period combined with the general pushing and shoving among local officials, the Control Board, and Congress to close off further consideration of a transnational migrant agenda. Other than Dave Clarke—for whom the civil rights battle was about redressing injustices committed by those with power against the powerless of every hue—the city's politicians generally ignored the concerns of the only growing segments of its population, Latinos and other "nonwhite/nonblack" groups. Marion Barry more closely captured the tenor of the time with his remarkable statement cited above that elected officials should be from the same "culture and race as the dominant population."[117]

Washington's post-1998 pragmatic municipal regime changed the local rules of the game in favor of Latinos in particular. In Jim Graham, the members of the Latino community at long last had a member of the D.C. Council who considered them to be a vital component of his core constituency. Greater attention to the nuts-and-bolts of making the city work opened up the possibility of political engagement with a Latino agenda.

This change has not been not welcomed by all. Dorothy Brazill—a longtime Columbia Heights activist, head of the citizens' group DC Watch, and a failed council candidate—made public statements that would have been unimaginable in many other American communities. When running unsuccessfully for the council, Brazill noted that the Latino community should face the fact that it does not count because "Washington is a black-and-white town."[118] In her view—as in Barry's reaction to the 1998 council elections—the city's political, social, and cultural life is legitimately defined solely by racial categories that emerged during decades of slavery, intolerance, and discrimination against Americans of African descent.

An increasing recognition on the part of the city's politicians that Latinos are a dynamic economic and potentially pivotal political force in municipal affairs only reinforced a local alliance system that sustained a D.C. regime favoring municipal pragmatism. Latinos pressed new concerns large and small; D.C. officials recognized those concerns as legitimate. Latinos consequently gained an interest in the present pattern of municipal politics. Issue by issue, Latinos and other nonnative Washingtonians slowly expanded the ground between what had long been separate white and black worlds.

The responses of present-day city officials to issues of particular interest to the transnational migrant community underscore this point. Foreign-born Washingtonians are making their presence felt in city politics for the first time by forcing policy outcomes and government decisions than would not have emerged without them and by raising issues that otherwise would have been ignored.

Not surprisingly, the provision of city services in languages other than English has been a paramount concern for many transnational migrant communities. The city, in fact, had established a program as early as 1976 requiring a number of city agencies to hire Hispanic managers so as to provide improved services for Spanish-speaking residents.[119] An Office of Hispanic Affairs was created within the D.C. Mayor's Office at that time, with an Office of Asian and Pacific Islander Affairs to follow, though both agencies would remain chronically underfunded for years.[120]

The D.C. public schools began to develop adult education curricula to enable Hispanics, Vietnamese, and other immigrant groups to learn English.[121] D.C. health administrators, increasingly concerned with the number of nonnative speakers of English showing up at public health facilities unable to communicate with medical personnel, began to develop interpreter programs.[122] Such efforts often received short shrift in a D.C. government overwhelmed by financial woes and reeling from the battles swirling around Mayors Barry and Kelly.

After 1998, the issue of multilingual language service provision became ever more prominent in D.C. Council deliberations, with Mayor Williams paying greater attention than his predecessors to the question. Council Member Graham introduced a bill in February 2003 requiring D.C. agencies that deal with the public to designate a language access coordinator. That official would be required to develop and to implement a plan that includes hiring bilingual employees and providing translated materials in those languages spoken by more than 3 percent of the D.C. population or by 500 individuals (whichever is smaller).[123] This act effectively requires two-dozen city departments (including those responsible for health, housing, education, motor vehicles, public works, corrections , aging, tax and revenues, as well as human rights) to offer services in Amharic, Chinese (Mandarin), Korean, Spanish, and Vietnamese.

Graham's proposal, initially supported by Williams and six council members,[124] was opposed by many in the D.C. bureaucracy who claimed that the cost would be prohibitive. Hearings on the bill often pitted agency officials testifying from the perspective of the previous civil rights municipal regime against council members who represented the city's increasingly pragmatic municipal regime.[125] The arguments of municipal bureaucrats evidently held little sway with the lawmakers, however, for the bill passed unanimously and was signed into law in the spring of 2004.[126]

By the end of that year, Aryan Rodriguez, the city's language-access director (a position created by the new law), told the *Washington Post* that the first eight agencies scheduled to be in compliance with the law by October 2006 had embraced the new policies.[127] Her primary challenges were to establish standard practices among agencies with quite different missions (e.g., the police department, public schools, fire and emergency services, and various health and human service agencies) and to find the funding required to carry out the mandate.

Laws and policies that require city agencies to provide services in the languages of local communities would hardly be remarkable in many an American city. Indeed, Arlington County just across the Potomac's Memorial Bridge in Virginia has long offered local services in several languages. What is striking about the Language Access Act is how much the city's population had changed before the issue could be introduced for consideration. Graham's bill revealed the time city's lag between demographic and social changes that had begun during the late 1980s and the political regime shift that occurred only in 1998, as well as a bureaucratic transition that is only now beginning.

## Services for Whom?

Discussions over how strictly to limit access by non-Washington residents to city services similarly exposed the varied time lines along which the city's society, political life, and municipal bureaucracy have been moving in recent years. The issue is especially acute in a metropolitan region where state lines crisscross a large number of relatively small jurisdictions. The effects of limiting access to services to local residents, while a long-held practice in many American cities, are magnified when populations are mobile, communities extend across jurisdictional lines, and family members live in loosely defined cross-generational extended households in which children and grandparents may move from one jurisdiction to another as they shift in and out among their children's, parents', and grandparents' homes. Metropolitan Washington's extensive and fluid Salvadoran, Korean, Vietnamese, Ethiopian communities—as well as those of several smaller migrant groups—have come to pose particularly difficult dilemmas for municipal service providers organized within self-contained bureaucratic and political units and mindsets.

The city's public health clinics and schools have become a focal point for a debate on the extent to which services should be preserved solely for its residents. If, say, an elderly Salvadoran woman living with her son in Mount Pleasant is being treated by a doctor at a city-supported clinic, what right if any does she have for continued care should she move in with her daughter across the river in Arlington? Jurisdictional and budgetary divisions so starkly drawn and understood by native-born Americans often appear as mutable to members of transnational migrant communities, for whom Western Avenue is merely a street rather than a boundary between two federal jurisdictions (Maryland and the District of Columbia). Organizations representing transnational migrant interests—such as the city's Counsejo de Agencias Latinas, the D.C. Kinship Care Coalition, and the Latin American Youth Center—increasingly demand that access to public services become more flexibly defined so as to reflect the metropolitan realities of residents' lives.[128]

Many D.C. politicians and residents are less sympathetic, especially in light of congressional prohibitions against so-called commuter taxes, which would be paid by those suburbanites who work in the city. In May 2003, Ward Six council member Sharon Ambrose proposed to heighten requirements for access to city services, especially in light of what she identified as a "steady stream" of children enrolling illegally in the city's schools without paying out-of-state tuition.[129] Ambrose's response—as well as the opposition to it from transnational migrant communities—highlights the extent to which Washingtonians born abroad now influence public policy discussions in the city. The assertion of the primacy of informal kinship networks over jurisdictional lines would have been unthinkable in an era when racially charged rhetoric dominated Washington and metropolitan life.

Mayor Williams demonstrated how much the city of Washington had evolved during his 2002 reelection campaign when he suggested that noncitizens living in the city should be allowed to vote in local elections.[130] Though unusual, noncitizens already have been extended voting rights in local elections in a number of communities across the United States, including in five municipalities in neighboring Maryland.[131]

The fact that provisions for legal noncitizen residents to participate in local elections is an accepted practice elsewhere in the United States did not stop national conservative commentators from pounding Williams for his proposal. Paul M. Weyrich, writing on the Web site of the self-described "conservative" Free Congress Foundation, drew an immediate connection from Williams to a point in the near future when millions of illegal immigrants would be stealing elections from rightful citizens throughout the country. "This is an idea that must be killed here and now," Weyrich wrote.

"It is bad enough," he continued, "that illegal aliens are accorded all sorts of benefits in some states, for which the taxpayers get the bill. To then give them the right to vote as well is just too much."[132]

Williams's initiative is unlikely to become law given congressional jurisdiction over the District of Columbia, as well as Republican control over both the House and Senate. Congressional approval of the 2005 D.C. budget, for example, was made explicitly contingent on voting rights *not* being extended to noncitizens.[133] Nevertheless, the act of placing such a proposal before the D.C. electorate is in and of itself an indication that transnational migrant communities are expanding their space to maneuver within the city's post-1998 pragmatic municipal regime.

Pragmatism similarly governed another change in the D.C. government's policies and practices in mid-2003. Tensions had been rising for some time between Latino as well as other transnational migrant communities and the Metropolitan Police Department as a consequence of local enforcement of various "homeland security" measures enacted following the terrorist attacks on New York and Washington on September 11, 2001. Federal insistence that local police compel observance of immigration laws in the normal pursuit of other law enforcement duties further eroded an already fragile partnership between D.C. police and local migrant communities.[134] Council Member Jim Graham had noted that such requirements increasingly meant that local transnational migrants were less likely to report crimes and cooperate with the police, given heightened fear that subsequent verification of immigration records would uncover grounds for deportation.[135]

Police Chief Charles H. Ramsey moved to alleviate fears within the transnational migrant community in July 2003 when, appearing with Council Member Graham, he announced that the city's police would no longer make inquiries about the legal residency of people contacted during routine police procedures.[136] Ramsey's policies were announced in Memorandum 84-41 issued by Mayor Williams and were confirmed in a Metropolitan Police Department Dispatch a few days later notifying D.C. police officers that inquires about immigration status were prohibited.[137]

Ramsey's actions, like Williams's proposal on voting rights and debates in the D.C. Council over service provision, reveal the extent to which the city's politics have evolved in pragmatic response to the arrival of thousands of new residents who were born outside the United States. Washington's city politics are no longer purely defined solely in black and white. Rather, new arriving residents who have never fit comfortably into such a starkly bipolar categorization are forcing local politicians and citizens to consider a middle ground that has been taking shape in all aspects of the city's life during the past generation.

## Creating a Middle Ground in the District of Columbia

Observers speak of Montreal as an intercultural city rooted in a French-language milieu. No one quite yet speaks of Washington as an intercultural city rooted in an African-American milieu. It is still not possible to speak of a cultural diversity in the American capital that embraces a locally dominant indigenous majority that is itself a national minority. Yet unlike a decade or so ago, such a statement is no longer unthinkable. The District of Columbia has expanded its capacities to accommodate diversity.

Other jurisdictions within the Washington metropolitan area are adjusting to the new realities being created by the arrival of transnational migrants with varying degrees of success. Arlington County, Virginia, is the most advanced in accommodating diversity, though there are numerous other examples of successful adaptation across the region. The evolution of the city of Washington is arguably more relevant to an exploration of how large urban communities can develop and implement "policies and institutions that have the overall effect of integrating diverse groups and cultural practices in a just and equitable fashion."[138] Washington has both a more troubled history of racial animosity and civil violence and also deeper scars remaining from the upheavals of the 1960s. The District of Columbia's peculiar constitutional status further complicates the task of creating diversity capital.

Demographic and economic—as well as political and policy—realities have edged Washington away from a bifurcated reality cast in black and white. The emergence in city politics of such issues as extending city services in languages other than English, recognizing the manner in which transnational migrant communities do not follow the region's political boundaries, giving noncitizens the right to vote in D.C. elections, and limiting police examination of immigration status point to a day when one might imagine Washington as an intercultural city in fact as well as in rhetoric. The pragmatism on display in recent city political battles, such as that over a baseball stadium, create the possibility for a middle ground to take shape between black and white. Latino Washingtonians, Asian Washingtonians, and Washingtonians from Africa and Europe alike are all reshaping their city into a more genuinely intercultural metropolitan center.

Intercultural though that city's future may well prove to be, Washington can not escape 200 years of race-based division. As Brian Ray has observed, diversity does not necessarily create social landscapes of equality and inclusion. The deep sociocultural and spatial divisions of race in Washington cannot be transcended merely by the presence of new transnational migrant communities.[139] The migrant presence must complement and become entwined with that of Washington's historic African-American community, acknowledging in meaningful ways the tremendous contribution of African-Americans to defining the Washington urban community and culture, to enhance the value of the city's and region's diversity capital. Transnational Washington must expand the total space of diversity for migrants, blacks and whites alike, and not merely somehow hover over the culture, economy, and politics of the region as if Africans and their descendants had not been participating in shaping it for more than three centuries.

The Washington story thus demonstrates the limits to the ability of communities to create new diversity capital as well as the opportunities for doing so. An intercultural Washington must remain rooted, in the end, in an African-American identity. As in Montreal, transnational migrants may create a middle ground between historically divided communities and realities. Again as in Montreal, they can do so by orienting their own cultures and needs within a context that is predicated on centuries-old communities that were in place long before the transnational migrants began to arrive.

About the time that Walter Fauntroy, Sterling Tucker, Marion Barry, Julius Hobson Sr., and other D.C. civil rights leaders were doing their best along the back roads of rural South Carolina to defeat Representative McMillan in his Democratic primary—thereby removing the leading congressional opponent of D.C. home rule—Kelly and Maze Tesfaye arrived in Washington from Addis Ababa to study biology and business.[140] Caught by changing political fortunes at home in Ethiopia, the sisters' few years of study turned into decades of life in the city of Washington.

Fifteen years after stepping off their plane in Washington, Kelly and Maze had accumulated a small nest egg from various modestly paying jobs.[141] Looking around, the sisters decided that they would be better off owning a restaurant rather than working as waitresses in the restaurants of others. Using all their savings, they invested in a small storefront at 5516 Colorado Avenue, Northwest, not very far to the east of the Carter Barron Amphitheater along the traditional racial boundary zone of Rock Creek Park.

The Tesfaye sisters had recalled along the way that they liked the people who hung around a jazz place where they once worked. Learning more and more about an African-based American art form—and thinking about the nice folks who played jazz—Kelly and Maze chose to convert their little restaurant into a club featuring jazz. From these modest beginnings would emerge two of Washington's best jazz venues—the original "Twins Lounge" on Colorado Avenue and, in the new millennium, "Twins Jazz," a second club on the upper floor of an old building along the reemerging entertainment strip at Fourteenth and U Streets, Northwest.

The original Twins Lounge opened on August 29, 1987, as racial tensions heightened with the arrest, trial, and imprisonment of Mayor Marion Barry. With Washington simultaneously heading toward the onslaught of a crack-cocaine epidemic and some of the highest murder rates in its history, the odds were long that the sisters would ever succeed in their little corner building reinforced by bars on every window and door. The sisters never let any of these events destroy their dream of a friendly tiny bar and restaurant in which everyone would be welcome. Eventually joined by the amiable photographer-bartender-host Joseph Beasley, the Tesfaye sisters made every customer—black, white, Asian, Hispanic, young, old, rich, poor, jazz lovers, and jazz novices—feel welcome. Walking into the original Twins was like stepping into someone's living room—only some of the best jazz talent in the world was at the piano.

Music lovers from all around discovered their little place, as did performers. Grammy winners who usually played at the prestigious Blues Alley over in Georgetown would seek out a chance to play "on their side of town" at Twins. On a hot summer's night, neighborhood locals might join city schoolteachers and White House advisers to hear this or that performer. Nothing mattered inside Twins except good music, good cheer, good fellowship, and the Tesfaye sisters' good Ethiopian food. Kelly and Maze had created the rarest of all rare spaces in Washington—one in which race did not matter.

Twins Lounge, which still exists at Fourteenth and Colorado, Northwest, and the newer Twins Jazz at Fourteenth and U Streets, Northwest, have enriched Washington. The two clubs preserve an increasingly rare opportunity for local and international jazz musicians to hang and to play. The music and fellowship are always more important than appearances and profits. Even at the nadir of racial hostilities around the time of Mayor Barry's conviction, Kelly and Maze preserved a warm civility that enveloped anyone who came to their little place.

Two immigrants from Africa have nurtured a cosmopolitan corner of Washington that is predicated on African-American culture (jazz). Just by being themselves, the Tesfaye sisters reveal to all who come to their little clubs just what Washington can become. Every night of every week of every month of every year for well over two decades, Kelly, Maze, and Joseph have been busily creating diversity capital for their new hometown.

Fragile though intercultural definitions of both Montreal and Washington appear at any given moment, those who care about both cities should recall the lethal communal violence that wracked these communities around 1970. Both North American cities and their metropolitan regions have required years before the wounds of the 1950s, 1960s, and 1970s could begin to heal. The possibility of an intercultural Montreal secured by a French-language core—and an intercultural Washington embedded in an African-American foundation—were unthinkable a generation ago. Culturally diverse migrants continue to change both cities fundamentally by successfully creating a middle ground that never existed previously. To recall the discussion in chapter 2, crossing Saint Lawrence Boulevard / Boulevard Saint-Laurent or Sixteenth Street, Northwest, no longer requires Montrealers and Washingtonians to place themselves in harm's way. Washington's stock of diversity capital—like that of Montreal—has increased.

If Washington has yet to secure the sort of comfort with diversity that Montreal has seemingly attained, Kyiv has only begun to come to terms with its new transnational migrant reality. The process of diversity capital creation is only just beginning in the Ukrainian capital.

# Chapter 6

# From Red to Orange:
# Kyiv's Post-Soviet Municipal Regime

Remarkable that on so fine a spring day the news should seem so amazingly peaceful. No shootings, no scandals. Quite the reverse, as if the paper was commanding its readers to be happy with life, with headlines inspiring gladness and hope.

—Andrey Kurkov, *Death and the Penguin,* 1996[1]

In politics, as in physics, low-probability events command attention. Only a few years before, a dispassionate observer would have concluded that the probability for a consummation of the quest for a quantum description of gravity would have been higher than the chance that the events which were about to transpire on Kyiv's main boulevard, the High Stalinist showcase Kreshchatyk, would ever take place.

The attention of thirty or perhaps forty thousand then-citizens of the Soviet Union was riveted on a quiet, theoretical physicist from the small western Ukrainian town of Lutsk in the Volyn' region. Climbing lampposts, clambering atop city buses, running every which way—thousands upon thousands of Kyivans watched in anticipation and disbelief as the young man walked purposefully out the doors of City Hall on a bright July afternoon. Tens of thousands of eyes latched onto his hand as he approached Kyiv's official flagpole.

Slowly the Soviet hammer-and-sickle came down, followed by a few fumbling movements. A spray of blue and yellow began to flow from the physicist's hands. With every hoist higher, the rising banner of an-as-yet-not-fully-born independent Ukraine unfolded into view. As the blue-and-yellow flag rose skyward, many in the crowd understood that they were no

153

longer Soviet. Instead, they were being rechristened as Ukrainians. His historic mission fulfilled, the physicist faded from view.[2]

Only a soothsayer of unusual prescience would have predicted Oleksandr Mosijyk's central role in the drama that was unfolding the streets of Kyiv throughout a broiling July 1990 afternoon. The thirty-five-year-old Mosijyk, like many others who came of age during the long rule of Soviet Communist Party general secretary Leonid Brezhnev, had studiously avoided politics as best he could for most of his life.[3] The local "brain," he left tiny Lutsk for Kyiv, where he revealed himself to be a promising theoretical physicist. The collective memories of a home town wracked by vicious repressions, bloody ethnic cleansing, and wicked civil strife during the imposition of Soviet, Nazi, and, yet again, renewed Soviet rule would travel with him wherever he went.

Entranced by the liberalizing opportunities unleashed by Mikhail Gorbachev's *perestroika* (restructuring) and *glasnost* (openness) campaigns, Oleksandr began to meet with small groups of liberal intellectuals around Ukraine, in Russia, and throughout the Baltic states. He eventually was drawn to a group of scientists forming around environmental activist Yuri Shcherbak.[4]

Shcherbak, a medical researcher who had been radicalized by the Chernobyl' accident, was increasingly involved in Gorbachev-era politics. He would serve in several distinguished posts, including as Ukraine's minister of environmental security as well as his country's ambassador to Israel and to the United States.

During the late 1980s, Shcherbak was just emerging as a significant presence on the Kyiv political scene after having published a devastating account of the spring 1986 accident at the Chernobyl' nuclear power plant.[5] His activism attracted numerous young scientists, such as Mosijyk, who banded together to sustain Shcherbak's victorious 1989 bid for a seat in the Congress of Peoples' Deputies in Moscow.

Flush from their candidate's victory, Mosijyk and other members of this circle concluded that Shcherbak's representation in Moscow should be accompanied by the additional presence of like-minded elected officials throughout Ukraine.[6] As part of this larger effort, Mosjiyuk and the electrical engineer Petro Kopil' agreed to stand for election to the Kyiv City Soviet (Council) in 1988 as a reform candidates from the Ukrainian Academy of Sciences. Their electoral bids proved successful, with both Mosijyk and Kopil' finding themselves among 620 Kyiv City Council deputies fighting to be heard in the behemoth municipal parliament which, true to Soviet tradition, had too many members and too little power to run a city.[7]

The Kyiv City Soviet—like the city councils of Moscow and Leningrad—became a battlefield between Communist regime loyalists and their anti-Soviet opponents. The Kyiv council divided evenly between the official Communists and reform-minded "democrats" (among whose ranks the articulate Mosijyk and Kopil' climbed rapidly). By mid-1990, council members had managed to reduce their assembly's membership to a still-unwieldy 300 members. The assembly remained as evenly divided as before, with a small group of centrists holding key swing votes on every issue.[8]

As decorum collapsed and more prominent politicians dove for cover, Mosijyk found himself serving as provisional chair of the Kyiv City Soviet Executive Committee (the *ispolkom*). Kopil' became his deputy, effectively making the two young scientists the city's acting mayor and acting deputy mayor.[9] Detecting rising nationalist sentiment among Communist officials—who were beginning to realize that they would enjoy more power as the governors of an independent Ukraine than as the overseers of a Muscovite satrap—Mosijyk used his moment at the center of local power to hoist the blue-and-yellow banner of a proposed independent Ukraine in front of City Hall.[10] On July 24, 1990, he walked into a growing throng of demonstrators outside City Hall and raised the banner of national sovereignty high above city's main avenue.

Exactly seventeen months would follow before the Soviet Union formally dissolved itself on December 26, 1991. During that time, Mosijyk was replaced as "mayor" by Grigorii Malashevs'kii, an engineer in the local state construction industry. Malashevs'kii would play a brief yet genuinely historic role in local politics.

Not having wanted to hold such a high ranking post, Malashevs'kii had been happy to head off to the Crimea for an August respite from the fights between the Executive Committee's "democrats" led by Mosijyk, and a Communist *nomenklatura* faction dominated by the former construction industry leader Oleksandr Omel'chenko.[11] Malashevs'kii arose early on August 19, 1991, to spend the morning on the beach. Aides found him there at 6:00 a.m. and informed him of an attempted coup against Soviet president Mikhail Gorbachev.

Malashevs'kii rushed back to Kyiv. Swayed by the words of the Lithuanian Communist Party leader Algirdas Brazauskas demanding a popular vote to name the Soviet president, Malashevs'kii began negotiations with Mosijyk, Omel'chenko, and military commanders to keep Red Army troops out of the tumultuous Ukrainian capital. He subsequently would recall his success in doing so to have been a crowning achievement of his career.[12]

Malashevs'kii soon stepped down, to be replaced in early 1992 by Kyiv State University Economics Faculty dean Vasil' Nesterenko.[13] Nesterenko established ties with two other professors-turned-post-Soviet mayors, Gavril Popov in Moscow and Anatole Sobchak in Saint Petersburg. Like Popov and Sobchak, Nesterenko was overwhelmed by the complexity of the economic collapse that accompanied the disappearance of the Soviet Union. Ivan Dan'kevich followed Nesterenko as "mayor" for just under three weeks in March 1992.[14] Eventually, Rukh activist Ivan Saliy proved able to consolidate his position as chief of the Kyiv State Administration with President Leonid Kravchuk who simultaneously appointed him to be the official presidential representative in the Ukrainian capital.[15]

## Kyiv's Post-Soviet "Soviet" Municipal Regime

Saliy's ascendency marked the beginning of a new regime in Kyiv municipal politics, one dominated by engineers and urban specialists who had once served Communist authorities. His appearance in conjunction with the establishment of a system of "presidential representatives" began to reorient local bureaucrats away from the rough-and-tumble politics of electoral competition and back toward the known predictability of bureaucratic rule. This change, at first subtle and seemingly benign, became ever more significant as Saliy and his successors established themselves in power. The sharp divisions within various city and district councils that had dominated the late perestroika era dissolved into a miasma of bureaucratic indeterminancy.[16] In retrospect, some observers came to view President Kravchuk's establishment of a system of appointed overseers as the moment when democratic politics within the Kyiv municipal administration began to fade away.

On a more personal level, Saliy had held a number of municipal positions before becoming Podil' district Communist Party secretary in 1979. By the late 1980s, he was head of the local Communist Party organization in his historic neighborhood. A transportation specialist by training and experience, Saliy moved quickly to catch the wave of reforms emanating from the Gorbachev-controlled Kremlin in Moscow—organizing street festivals, fighting to upgrade historic streetscapes, and working generally to turn Podil' into an increasingly lively and congenial intown district.

Saliy became ever more outspoken in his support for Ukrainian independence, sponsored the reconstitution of the centuries-old University of Kyiv–Mohyla Academy, and joined the ranks of municipal reformers such as Mosijyk, Kopil', and Nesterenko. Unlike his more idealistic former-academic colleagues, Saliy-the-transportation-specialist understood how to make a city work. Moreover, he enjoyed the continuing support of Ukrainian president Kravchuk.

The Ukrainian weekly *Den'* tellingly captured Saliy's contradictory attributes and in doing so effectively characterized many of those who were now members of the city's postindependence municipal elite. "Ivan Saliy," the paper wrote in 1998, " belongs to the perestroika generation of politicians. He can hardly be identified as a democrat or reformer, but he was also not typical of the party *nomenklatura* of the period. He did his job well, and the results were obvious in the restoration of the Podil' section of Kyiv. All this is most likely the result of his character, not outlook or ideology."[17]

Leonid Kuchma's defeat of Kravchuk in Ukraine's 1994 presidential elections removed Saliy's chief patron from power. The new president named Leonid Kosakyvs'kii to replace Saliy as presidential representative and city chief. Kosakyvs'kii had held a number of district positions in the Pechersk District of central Kyiv, where he once had worked as a manager at the Arsenal defense plant. Subsequently, he served in several positions within local Communist Party organizations. Kosakyvs'kii's tenure marked the further return of old Soviet-era municipal officials to power in Kyiv and the concomitant growing isolation of leaders of the democratic movement such as Oleksandr Mosijyk.[18]

Holding an increasingly weak hand, Mosijyk knew that his days as in municipal politics would soon be over. He moved to banking, and then on to found a liberal-minded think tank and to organize political parties. He would serve in the short-lived reform administration of Prime Minister Viktor Yushchenko. Mosijuk eventually joined the brain trust promoting the oppositionist Our Ukraine Bloc that would foster the so-called Orange Revolution in November and December 2004.

Back in 1996, before he abandoned his involvement in municipal affairs, Mosijyk was searching for a compromise candidate who could bring order to city affairs. Before leaving city government, Mosijyk proposed that a skillful member of the local construction industry be appointed to lead Kyiv.[19] President Kuchma followed this advice, naming Mosijyk's preferred candidate, Oleksandr Omelchenko, as Kyiv mayor on August 12, 1996.[20]

Kopil', sensing the same pressures, initially sought refuge in the one city district that had retained a reform-oriented administration and council, the Moskovskii *raion* (district).[21] He would hold a number of positions on the district council until Mayor Omelchenko abolished this last bastion of reform in a process of district consolidation that reduced the total number of submunicipal regions within Kyiv from fourteen to ten. Mayor Omelchenko simultaneously used Kopil's sixtieth birthday as an excuse to force his once stalwart supporter into retirement from local politics.

Omelchenko, for his part, garnered additional support from the local *nomenklatura* as one of their own. Long a dominant force on the Kyiv City Council, Omelchenko had worked for years in the local construction industry, as well as in Afghanistan during the late 1980s as a manager of large-scale construction projects. Reformers such as Mosijyk and Kopil' viewed the new mayor as an experienced manager practiced in the art of compromise.[22]

Omelchenko's political base embraced pensioners, Afghan war veterans, Chernobyl victims, and human rights activists who recalled Omelchenko's support of dissident physicist Andrei Sakharov during early perestroika days. As a representative of an old Kyiv family and a local "patriot," Omelchenko's desire to rebuild local monuments destroyed by the Bolsheviks made his appointment attractive to local nationalists and democrats.[23]

Because he was an effective administrator, Omelchenko's popularity rose rapidly among Kyiv city residents once he was ensconced in power. He handily defeated his opponents in the city's first general election for mayor in August 1999.[24] The mayor opened his administration to many different groups, appointing his onetime rival Ivan Saliy as deputy mayor in charge of transportation services. Always a gifted and innovative manager, Saliy won kudos while serving Omelchenko by upgrading metro service and launching a fleet of ubiquitous bright-yellow minibuses on routes throughout the city.[25]

Omelchenko craftily used construction contracts to extend his reach over a massive city bureaucracy. His administration undertook numerous high-visibility renovation projects—such as repaving the wide sidewalks along the city's famous central artery, the Krestchatyk, with gray brick. The mayor's critics viewed such efforts as giant money-laundering schemes for Omelchenko's own family, friends, and corrupt associates. Omelchenko's supporters argued in return that Kyiv never looked better.

Many Kyivans came to appreciate their mayor's candor about the city's problems. For example, in response to a question posed during a 2000 interview about aspects of the city infrastructure that do not meet European standards, the mayor replied, "Thank God, our people are inherently healthy individuals and our medics are so skilled that they can perform the most complicated surgeries, figuratively speaking, with a pair of scissors and an ax."[26] By the time he stood for reelection in 2002, Omelchenko was able to claim 72 percent of all the votes cast.[27]

Omelchenko's time in office proved to be a period during which local elites steadily consolidated a new post-Soviet "Soviet" municipal regime. Reformers were swept away by a Ukrainian Communist Party elite, the *nomenklatura*, which began to detect distinct advantages in running their own country rather than merely dancing to instructions emanating from Moscow. Many of the people and institutions that had once ruled over Soviet Ukraine remained in place (or cosmetically reinvented themselves).

None other than Ivan Saliy captured the essence of this new municipal order in an especially frank 1998 interview with reporters from the popular newspaper *Den'*. "I think that, no matter how this elite may change," Saliy began, "it cannot adequately assess the situation in and outside Ukraine at this transitory stage. These people turned out to be professionally unprepared, and lack of professionalism leads to the loss of patriotism and an inability to formulate and protect national interests. Unless we change our attitude to what we call cadre policy, nothing will change in the government machine and society. When morals are discarded and not replaced, society starts to fall apart."[28]

## Soviet Rules of the Game for Ukrainian Post-Soviet Law and Practice

The operation of Soviet cities had evolved by the 1980s into a complex revel among Communist Party officials, city administrators, and central ministries. According to the 1977 Soviet (so-called Brezhnev) Constitution and its accompanying legislative revisions, formal power lay with city "*soviets*"—very large "elected" councils with symbolic presence at best.[29] The soviets, in turn, selected an "executive committee" (the *ispolkom*) which had responsibility for the daily management of municipal affairs. The executive committees operated a local municipal administrative system organized around functional departments charged with picking up the garbage, paving streets, building and managing housing, running shops for basic consumer goods, and the like. The physical development of a city such as Kyiv rested with the local city planning committee attached to such an executive committee.

Branches of the central ministries and their factories and enterprises ran their own urban services from housing to bus systems, from bread stores to clothing outlets. Given the autarky of the Soviet economy, the ministries and enterprises did their best to control as many goods and services as possible on their own in the hope that they could guarantee at lease minimal quality and predictable distribution. Ministerial branches dominated many Soviet cities, though they were forced to compete for pride of place in a republican capital such as Kyiv. Far too many claimants for resources roamed Soviet-era Kyiv for any single federal or republican-level ministry to lord over the allocation process for long. The ministries, in turn, were subordinate to state planning agencies (Gosplan) and, eventually, the Council of Ministers.

All these competing fiefdoms fell under the supervision of a set of Communist Party committees, which, though they enjoyed little formal constitutional authority, nonetheless dominated the city through control of personnel appointments (the *nomenklatura* system). Not only were individual officials subject to internal Communist Party discipline, but local party officials also could send signals up their own chain of command and control through regional, republican, and national party committees to transmit demands back to their local partners through ministerial and municipal chains of command.

This system proved remarkably resilient despite the collapse of the Communist Party.[30] Individual administrative districts—the *raiony*—collected national and local taxes within their jurisdiction and forwarded such revenues to higher authorities. City officials redistributed those taxes collected in Kyiv among municipal agencies and departments, as well as the city's ten *raiony*. Usually about 7 percent of the total revenues originating in a given district were returned to that district by the time this process had been completed. District officials often complained bitterly about the arbitrary nature of this process, because the budgetary transfers were calculated on the basis of long-established revenue-sharing formulas.

The Soviet-style system of revenue collection accentuated entrenched perceptions within lower administrative levels of municipal administration that city and district officials were accountable first and foremost to bureaucratic superiors rather than to local citizens. Such views ran counter to patterns of revenue collection, in which a Soviet-era reliance on enterprise taxes was giving way to an increasing postindependence emphasis on taxes levied on individual citizens. By 2000, individual residents had come to constitute the largest single category of taxpayers in most Kyiv districts.[31] Moreover, the services provided by the city's districts—which included schools and preschools, health care, culture and sports, a portion of housing, partial maintenance of roads, and general landscaping—arguably affected the quality of life of individual residents more than the effectiveness of local enterprises.[32] Taxpayers around the city were increasingly aware of their growing role in paying for municipal services, even as many city administrators continued to orient themselves toward the demands of higher-level municipal officials.

The Soviet system's legacies were conceptual as well as structural, being shaped by how local elites thought about the role of local government and statecraft. Large construction trusts (developers who made decisions on the basis of bureaucratic expediency rather than the possibility of profit) dominated formal municipal structures within the world of the *ispolkomy*. Larger strategic thinking, to the extent that it existed at all, rested within the Communist party bureaucracy; economic planning, such as it was, focused on ministerial-based organizations. The collapse of Soviet power, therefore, left the daily administration of Kyiv and dozens of other cities and towns in the hands of municipal officials who thought about community management as a construction challenge rather than as a political, economic, or social management task. Given these earlier patterns of thought and action, local Soviet-era and their postindependence successor bureaucracies were as unprepared for approaching their cities as social organisms as they were for the introduction of market economic principles.

At the same time, Ukraine had moved farther than other post-Soviet states toward establishing formal municipal legal and administrative structures predicated on Western principles. A consensus emerged around the proposition that the Ukrainian system of local administration should draw as much as possible on European models of civic management.[33] A February 4, 1994, law on local administration achieved many of these goals by modernizing municipal finance and administration for most of Ukraine.[34]

National officials continued to squabble over Kyiv's status as the national capital. Pursuant to articles in the Constitution of Ukraine that granted the capital special legal status, separate legislation on Kyiv as a capital city became law in January 1999. As in Washington, Kyiv officials became subject to the control of their national government to a far greater degree than their counterparts elsewhere.[35] Significant juridical ambiguity and legislative imprecision remained. For example, the "mayor" of Kyiv was both elected by popular franchise as chair of the Kyiv City Council and also appointed by the president of Ukraine to be chief of the Kyiv City Administration. The absence of legislative clarity, typical of Soviet-era laws, unleashed an unending wave of conflict among Kyiv municipal and Ukrainian national authorities.

Following Omelchenko's 1999 electoral victory, President Kuchma appointed the mayor to serve simultaneously as the chairman of the Kyiv State City Administration.[36] As the months passed, Omelchenko's popularity continued to rise even as Kuchma was becoming ever more embattled in the wake of allegations surrounding the murder of the journalist Heorhii Honhadze. In 2002, President Kuchma recommended that all civil servants who were running for seats in that year's parliamentary elections should take a leave of absence from their state duties throughout the campaign. Most Ukrainian politicians ignored their president. But Omelchenko honored the request and petitioned for vacation leave from February 11 until election day on March 31. He appointed Kyiv City Council secretary Volodymyr Yalovyi to be acting mayor, naming deputy Kyiv State City administrator Mykhailo Holytsya as acting chair of the city administration.

Kuchma accepted Omelchenko's leave request, but he appointed another city deputy administrator tied to Prime Minister Anatoly Kinakh, Ihor Shovkun, to become the city administration's acting chair during Omelchenko's "absence." Omelchenko immediately took to the airwaves of the American-supported Radio Liberty to protest, touching off public denunciations among local politicians over who was in charge of the Ukrainian capital's administrative structure.[37] Kuchma, sensing a public relations disaster, backpedaled. He canceled his initial order, thereby restoring Omelchenko's authority over the Kyiv City Administration during the parliamentary campaign.[38]

The issue was not resolved, however, because the root cause of the contretemps—Omelchenko's continuing popularity and Kuchma's collapsing power base—remained unaddressed. Kuchma seemingly bided his time, waiting for Omelchenko to reach the mandatory retirement age for Ukrainian state employees of sixty-five years of age (which the mayor would inevitably reach during the summer of 2003). Not able to undo Omelchenko's popular mandate as "mayor," Kuchma and his cronies simply declared the mayor ineligible for his post as chief of the city administration and appointed Valeriy Khoroshkovskyi, Kuchma's minister of economy and European integration, in Omelchenko's stead.[39] The mayor successfully beat back Kuchma's renewed attempt to force him from office.[40]

Just as Omelchenko appeared to weather his latest storm, Parliamentary Deputy Leonid Chernovetsky, from the city's Darnytsia district, filed an ultimately unsuccessful suit to have Omelchenko removed from his elected post as mayor, citing the recall of California governor Gray Davis as a precedent.[41] Omelchenko steadfastly refused to give up the reigns of municipal power, continuing to manage his city seemingly unperturbed by the allegations flying all around him. Similar battles erupted in cities across Ukraine as contending groups sought to seize control over the tax funds flowing through municipal coffers for their own personal gain.[42]

These unhappy tales of municipal intrigue and greed revealed a continuing hold of deep and highly destructive post-Soviet "Soviet" municipal regimes over Kyiv and numerous other Ukrainian cities. Municipal budgets became viewed as sources of income to be diverted for private purpose; the primary function of city administration was seen first and foremost to be that of "builder" (construction being another source of income for well-connected former *nomenklaturshchiki*); and key founding legislation and city charters were purposefully equivocal, with social welfare policies—and the urban poor—being abandoned in the name of adherence to macroeconomic discipline.[43]

Well over a decade after Ukrainian independence, Kyiv's city administration had become arguably less capable of coping with grand issues of urban management than when idealistic Oleksandr Mosijyk stepped from City Hall on a hot July day in 1990 with a blue-and-yellow flag in his hands intent on declaring Kyivans to be "Ukrainians" rather than "Soviets." Meanwhile, tens of thousands of migrants had arrived from beyond the borders of Ukraine, presenting city administrators with challenges unimagined when Kyiv was still a fully Soviet city.

## Enter Transnational Migrants

This brief overview of Kyiv's recent history is intended to establish some parameters for comparison with Montreal and Washington. Kyiv has experienced widespread communal strife, which reflected deep divisions within the city. Conflicts over the meaning of Ukrainian nationalism have been bitter, as was the case with Quebec nationalism in Montreal. As with African-Americans in Washington and francophones in Montreal, the local majority—in this instance ethnic Ukrainians—constituted an embittered minority within a larger state—the Soviet Union. Large-scale public demonstrations closed down the normal functioning of the city for some time, leading in Kyiv to a collapse of the entire political system in late 1991. Kyiv's expanding community of transnational migrants—like those in Montreal and Washington—remained relatively unconcerned with the questions of identity and language that had so dominated Kyivan life in the past.

The pacing of events in Kyiv has been different, with the Soviet regime long repressing overt expressions of discontent. Consequently, Kyiv may be at an earlier stage in the development of a cycle favoring the creation of diversity capital than is evident in Montreal and Washington. Alternatively, the city may be on a different trajectory altogether. These points of similarity and contrast present a felicitous opportunity for comparison among all three cities.

Many favorable conditions exist in Kyiv for the creation and accumulation of new diversity capital. For example, the city's migrants generally retain a positive attitude toward both Kyiv and Ukraine. Nearly all migrants speak Russian, many have learned Ukrainian, and most migrant families enroll their children in local schools—thereby demonstrating a basic commitment to life in the Ukrainian capital.

These attitudes were confirmed in a survey of Kyiv's migrant communities carried out by the Kennan Institute's Kennan Kyiv Project during mid-2001.[44] According to the results of this survey, vast majorities of all groups (except for Arab migrants) highly valued the absence of war in Kyiv. Moreover, they appeared ready to trade peace for a multitude of inconvenience.[45]

Such favorable attitudes underlie the willingness of the majority of migrant respondents—with the important exception of African migrants—to declare that they would have moved to Kyiv even had they known all that would happen to them in the process (including 89 percent of Vietnamese respondents to the Kennan Migrant Survey Project, 65 percent of Arab migrants, 58 percent of Kurdish migrants, 48 percent of Pakistani migrants, and 46 percent of Afghan migrants). These figures contrasted sharply to the 88 percent of African respondents to the Kennan Institute survey who would not have come to Kyiv had they known what experiences were in store.[46]

A more complex picture emerged when migrants were asked whether or not they would like to leave Ukraine.[47] More than 70 percent of African and Afghan respondents to the Kennan Institute survey declared their general intention to move elsewhere at some indefinite time in the future, as did a majority of Chinese and Pakistani survey participants (in contrast to fewer than half of Vietnamese, Chinese, Arab, and Kurdish sample members).

Nearly all migrants who stated an intention to leave identified their preferred new home to be in a Western country. However, fewer than 10 percent of Afghans, Vietnamese, Africans, and Pakistanis reported specific plans to depart within the next year (together with 16 percent of Arab and Kurdish respondents, as well as 26 percent of Chinese survey participants).

This difference between an inchoate inclination to move elsewhere and specific travel plans was explained in part by the fact that most Afghan, Arab, and Kurdish migrants expressed a desire to receive Ukrainian citizenship.[48] Even a majority of disgruntled African survey respondents who complained bitterly about racial discrimination and animosity reported that they intended to seek Ukrainian passports.[49] Once again, some preconditions appeared to exist for the creation of diversity capital, as reflected in the institutional capacity of the Ukrainian state to incorporate transnational migrants into its community of citizens.

To varying degrees, then, Kyiv's migrant communities appeared to consist of city residents whose primary concerns—and demands on local authorities—were of an immediate and pragmatic nature. Ukrainian sovereignty was taken as a given for transnational migrants (the country already having become independent before the arrival of the preponderance of the city's transnational migrant residents). Moreover, the large-scale infrastructure projects of such intense interest to a local government dominated by officials with ties to the construction industry were hardly salient—or profitable—to communities still thinking about moving on. Migrant disinterest in local construction industries was heightened further by their exclusion from access to the consequent economic benefits so jealously guarded by those with ties to City Hall.

Generally, Kyiv's migrants proved themselves to be autonomous economic actors who were able to capitalize on their economic success to secure housing and material goods for themselves and their families largely outside the official state economy.[50] Migrants nonetheless were not totally self-reliant. They often required the assistance of government officials and agencies in three areas that depend of the performance of official institutions: health care, education, and police protection.[51]

These three policy areas were often of little interest to officials nurtured in Kyiv's post-Soviet "Soviet" municipal regime. Few city officials tied to Mayor Omelchenko's patronage networks evinced much interest in social policy, either for migrants or for native-born Kyivans. Municipal agencies so heavily dominated by construction industries lacked the resources and capacity to adequately address the social needs of native-born local residents, let alone of migrants from abroad. When asked about the presence of transnational migrants and their needs, officials in City Hall either denied the presence of foreign residents or responded that Ukrainian citizens were suffering as well.[52]

The absence of municipal workers assigned to work on migrant issues provided evidence of a deficient official capacity to deal with the presence of a significant community of transnational migrants in Kyiv. As late as 2001, a time when new citizenship laws were being implemented that would encourage the already tens of thousands of transnational migrants living in Kyiv to think of securing Ukrainian passports, the city administration's Department for Refugees and Minorities employed just two case workers. The salaries for those employees were paid for by funds drawn from the municipal budget, plus an additional two case workers whose salaries were covered by the Office of the United Nations High Commissioner for Refugees.[53] Not surprisingly, given such startling inattention to their needs, migrants were becoming increasingly disgruntled with municipal authorities just as municipal officials remained deaf to the concerns of migrants.[54]

By mid-2004, Kyiv appeared to be at the moment of policy realization at which Washington and Montreal found themselves a generation or so ago. Thousands of migrants had arrived in the city from abroad, largely unnoticed by the local policy community. City officials and the broad public remained preoccupied with recurring policy concerns that had long dominated the local scene. A new public policy agenda connected to the appearance of the city's transnational migrant community was taking shape beyond the reach of traditional elites. All that remained before further action would be possible was to pass through the presidential elections scheduled for Halloween of that year.

## Painting the Town Orange

Thousands of protesters took up residence on Kyiv's central Maydan Nezalezhnosti (Independence Square) and on its premier street, the Khreshchtyk, throughout November and December 2004 and into January 2005. Their numbers swelled into the tens and even hundreds of thousands at times, perhaps surpassing 1.5 million at pivotal moments over the course of these cold winter weeks. The demonstrations and negotiations unfolding in the cities streets and behind closed doors followed massive election fraud in the November 21, 2004, presidential runoff. They would lead eventually to the annulment of the disputed election results and the ratification of far-reaching constitutional reforms. These events became known as Ukraine's Orange Revolution.

The election campaign had reached full fury as summer came to an end, by which time twenty-six candidates were registered officially to run for the post of president of Ukraine[55] The contest narrowed quickly to two large opposing political positions represented by the candidacies of Prime Minister Viktor Yanukovich and former prime minister Viktor Yushchenko. Both men emerged as symbols for deep philosophical divisions within Ukraine.

Yanukovich, and his "party of power" stood for a continuation of the status quo; for sustaining a "patrimonial" approach to governance through which the state writ large would remain as the central actor in Ukrainian life. Citizens, in this view, were to subordinate their interests to those of the state. Supporters of this position felt that their country's best interests would be served through closer contact with Russia. Yanukovich, according to Ukrainian public intellectual Mykola Riabchuk, represented a "subject" political culture.[56]

Yushchenko, conversely, stood for change and a more "modern" approach to governance whereby the state would serve citizens, who would themselves become the primary autonomous actors within society. He—and his campaign's official color of orange—came to symbolize a Ukraine that would seek closer ties with the European Union. Yushchenko, according to Riabchuk, represented a Ukraine that would become a nation of "citizens."[57]

The unexpected disappearance of Yushchenko in early September amplified the already intense conflict between these two mutually exclusive worldviews. As events unfolded, Ukraine and the world would come to know that Yushchenko had suffered from dioxin poisoning. Concerned aides saved his life by rushing him to the Rudolfinerhaus Hospital outside Vienna. Yushchenko remained hospitalized there for eight days before checking out against the advice of his doctors. He reappeared in Ukraine horribly disfigured and crippled by pain.[58]

Yushchenko's face of horror became a ready-made symbol for those Ukrainians who had had enough of what they believed to be a thoroughly corrupt regime. The grotesque lesions and ulcers that contorted his once handsome face combined with a tube poking out from his back through which medication could administered directly into his spinal canal—ready symbols for everyone of just how high the stakes had become.

Yanukovich similarly reminded everyone of the civilizational battle being played out across Ukraine just by being himself. The tall, heavyset "Don of the Don," Yanukovich was the former leader of the Donbass Region, once among the Soviet Union's most powerful industrial heartlands. The area reputedly fell prey to criminal overlords at the outset of Ukrainian independence and seemed to function as a world unto itself. Yanukovich was himself a convicted felon—having served jail time for robbery as a teenager and subsequently for an assault charge. He spoke a Russian sprinkled with prison slang, and a Ukrainian that barely resembled the language of educated elites in Kyiv.[59] These audible traits combined with the smothering embrace of Russian president Vladimir Putin to cause great consternation among Yushchenko supporters while making Yanukovich "our guy" (*nash chelovek*) to hundreds of thousands of Ukrainians who longed for the "good old days" before independence.[60]

The election's first round ended inconclusively in already troubled balloting on October 31. Yushchenko eventually was declared the front-runner in a vote count that lasted days (and which had initially be announced as having tilted in Yanukovich's favor). According to the final official vote tally, Yushchenko had polled 39.87 percent of the voters that Sunday, as opposed to Yanukovich's count of 39.32 percent of all the votes cast. One way or another, a runoff was necessary. The stage was set for the decisive final round to be held on Sunday, November 21.[61]

As the polls were closing that evening, international observers released exit polls projecting a Yushchenko victory. Forecasts based on these surveys ranged from a 54 to 45 percent Yushchenko victory to a 49.4 to 45.9 percent spread in Yushchenko's favor. Widespread rumors of fraud were surfacing even before the last votes had been cast. Consequently, Yushchenko's supporters were primed to believe that their candidate had won even before the official results were released. From their perspective, any outcome other than an orange victory would be the result of massive fraud.[62]

The Central Election Commission announced quite different results, placing Yanukovich in the lead at 49.4 percent to Yushchenko's 46.7 percent.[63] Widespread statistical anomalies as well as a cavalcade of reports concerning intimidation and shenanigans of various kinds immediately called the result into doubt. Before the evening was out, the Organization for Security and Cooperation in Europe's Office for Democratic Institutions and Human Rights released a statement concluding: "As in the first round, state executive authorities and the Central Election Commission displayed a lack of will to conduct a genuine democratic election process."[64] By the time the commission had certified their official tally, Yanukovich was reported to have received 49.46 percent to Yushchenko's 46.61 percent.[65]

Not surprisingly, in a city where more than 70 percent of the voters had sided with Yushchenko, crowds of protesters immediately ran from their apartments into Kyiv's subfreezing streets. By Monday morning, thousands of disgusted citizens under the leadership of the student movement Pora (It's Time) had taken up residence in a tent city that spread throughout Maydan Nezalezhnosti along the Khreshchatyk just as tense political battles were beginning in the Verkhovnaia Rada (Ukraine's Parliament), well within shouting distance just up a steep hill from the east end of the Khreshchatyk.[66] A number of city councils across the country—including that of Kyiv—declared the election results invalid; Mayor Omelchenko announced that city authorities would not try to clear the city's streets of protesters.[67] Meanwhile, Russian president Putin was already congratulating Yanukovich, and pro-Yanoklovich city and regional administrations in southern and eastern Ukraine were threatening secession should their candidate be prevented from assuming power.[68]

The stage was set by the early hours of November 22 for weeks of intense behind-the-scenes negotiations involving senior European Union, Polish, Lithuanian, Russian, and Ukrainian officials; protracted legal battles; stormy parliamentary debates; opposing claims of victory; and secession threats by pro-Yanukovich districts in the east (e.g., Donetsk and Kharkiv). This high drama was played out before throngs of hundreds of thousands of people (reaching well over a million at times) coursing through Kyiv's streets.[69]

These events were far too intricate to record here. The turning point in the story proved to be a dramatic hearing of complaints concerning election fraud brought by the Yushchenko campaign to the Ukrainian Supreme Court. That session, which commenced on November 29, was carried out in full public view and broadcast throughout Ukraine on television. The Court's hearings took place against the backdrop of continuing unrest, threats of various kinds, and political wheeling and dealing. Everyone seemed to jump into the act at one point or another, with high-level negotiations chaired by Javier Solana, the European Union's high representative for the common foreign and security policy, drawing into the same room the likes of Alexander Kwasniewski, the president of Poland; Valdas Adamkus, the president of Lithuania; Boris Gryzlov, the chairman of the Russian State Duma; President Kuchma; Prime Minister Yanukovich; Viktor Yushchenko; and Volodymyr Lytvyn, the chairman of the Ukrainian Parliament.[70]

In the end, the justices of the Ukrainian Supreme Court would have the decisive word. In a decision released early on the evening of December 3, the Court ruled that members of the Central Election Commission "did not investigate the reports of the territorial elections commissions concerning the voting process within the boundaries of their electoral territories, and did not check their authenticity, their truthfulness, or their completeness."[71] The justices continued: "At the time the CEC [Central Election Commission] delivered its final results on the runoff vote for the presidency, the courts had not yet completed their examination of the complains submitted concerning the inactions, actions, and decisions of the territorial electoral commissions."[72] The actions were therefore "not in accord with law. Consequently, the decision adopted by the CEC is void."[73]

Having annulled the results of the November 21 runoff, the Supreme Court continued on to cite a number of violations of law and legal principle that rendered the electoral process unfree and unfair. The justices rejected the proposal of Yushchenko's lawyers to certify their candidate as the victor. Instead, they scheduled a new nationwide runoff, which came to be held on December 26. "In view of the impossibility of establishing the true results of the will of the electorate in the country-wide balloting by means of tallying the votes in the runoff poll," the Court concluded, "and, in view of the fact that the November 21 runoff poll did not change the status of the candidates the Court holds that it is necessary to renew the rights of the subject of the electoral process by holding a runoff poll, in accordance with the rules set down in the presidential election law."[74]

Political negotiation among Ukraine's power elite became increasingly intense as the joyous demonstrators on the Maydan celebrated their breakthrough moment.[75] Despite some additional false starts and breakdowns in comity, the advantaged clearly swung to favor the Yanukovich forces.[76] A final compromise deal passed Parliament on December 8, including a package of laws setting the new runoff date, changing the election law (the Constitutional Court would subsequently overturn some of these provisions), and approving constitutional provisions that had long been favored by President Kuchma to transform Ukraine from a presidential to a Parliament-dominated form of government in 2006.[77] Ukraine, much to the jubilation on the Maydan, was set on a new course.

The lead-up to the December 26 vote proved to be almost anticlimactic. Much (but hardly all) of the crowd on the Maydan and Khreshchatyk dispersed, confident of a Yushchenko victory. A nationally televised debate revealed little beyond what was already known to all.[78] A total of 37,291,681

Ukrainians went to their polling places for a third time in two months to elect Viktor Yushchenko their new president by a vote of 52 to 44 percent.[79] As in the previous ballots, Yushchenko carried sixteen regions and Kyiv City to the north and west of the country (78 percent of Kyivans cast their ballots for Yushchenko), while Yanukovich piled up equally impressive totals in nine regions and Sevastopol City to the south and east.[80] Yushchenko was inaugurated on January 23, 2005, after a few more weeks of unsuccessful legal maneuvering by Yanukovich.[81]

Although there appears to have been an air of inevitability about the success to the Orange Revolution in hindsight, any number of alternative outcomes were possible, especially had violence broken out. The fact that the crowds of demonstrators were extremely well disciplined helped to ensure a peaceful outcome. Moreover, many in the security forces visibly backed away from the use of force very early on in the demonstrations. For example, the SBU (the national security service, the Ukrainian descendant of the Soviet KGB) head General Olexander Skipal'skiy and four other officers joined in the pro-Yushchenko demonstrations on November 25, appealing for law enforcement agencies to be with the people.[82]

Subsequent reports reveal that the Interior Ministry was prepared to resort to force. The ministry went so far as to send its forces into the crowded streets with live bullets on Sunday, November 28.[83] The security forces, like so many institutions in the country at the time, were divided within themselves.

Most important for the outcome of the Orange Revolution, people on all sides of the barricades understood the horrid conflagration that surely would have followed in the wake of any violence.[84] In the end, peace, politics, compromise, and tolerance prevailed.

The Orange Revolution was without question a defining moment in Ukrainian history, and an event of international importance. As with the 1970 October Crisis in Montreal—and the 1968 Martin Luther King Jr. assassination riots in Washington—Ukraine's Orange Revolution was indisputably a local Kyiv event as well. What happened on the city's streets during November and December 2004 both reconfigured the future of the Ukrainian capital as an urban community and, in very real ways, was a product of the city itself.

Kyiv literally shaped events, first by its physical form.[85] The city's main street—the Khreshchatyk—was rebuilt as one of the Soviet Union's premier examples of Stalinist urban planning following its total destruction during World War II. The avenue arose refreshed by huge, florid examples of totalitarian architecture matched in scale and extent only in parts of Moscow and East Berlin. Approximately one-third of the way along its route from what was once Komsomol (Young Communist League) Square in the east to Bessarabskaya Square in the west, the Khreshchatyk explodes outward into a large square to be used for appropriately grand official Communist Party demonstrations.

This square was renamed Maydan Nezalezhnosti, or Independence Square, following 1991. Mayor Omelchenko sponsored the construction of underground shopping malls along the Khreshatyk both at Maydan Nezalezhnosti and under Bessarabskaya Square. President Kuchma and Mayor Omelchenko oversaw a simultaneous tacky tarting up of the aboveground areas to mark the tenth anniversary of Ukrainian Independence in 2001.

Over time, Omelchenko's city government built a "temporary" stage for rock concerts that was outfitted with stadium-size television screens and sound systems. The mayor sponsored closing the Khreshchatyk and the Maydan to vehicular traffic on Sundays, creating an enormous outdoor space for promenading. Kyivans adopted the entire area as their own, with upward of half-a-million people strolling about, shopping and listening to music on any given Sunday.

The Maydan was a perfect location for such a central public space. Nestled in a small valley among various fragments of the overall city (Pechersk, "Old Kyiv," Bessarabka, etc.), the Maydan exerts a central gravitational force giving form and definition to Kyiv's urban life. As many as a dozen streets flow down into the Maydan and the Khreshchatyk from various angles. Nearly all the city's major institutions are located nearby. The Parliament (Verkhovna Rada), the Presidential Administration, Saint Sophia's Cathedral, the offices of the Central Election Commission, City Hall, and the headquarters building of Ukraine's trade unions are all within a brisk fifteen-minute walk of Omelchenko's sound stage. Apartment houses first built for prominent members of the Soviet regime similarly are close by, many occupied by residents ready to cook a warm meal to feed demonstrators camped on their doorsteps. Because it is connected to the entire city by several subway and major bus lines, the Maydan had become the focal point for civic life well before demonstrators turned it orange. Protesters naturally headed straight for the Maydan on the night of November 21–22 because it appeared that someone was trying to steal the country's presidential election.

C. J. Chivers, who covered the Orange Revolution for the *New York Times,* observed that many factors sustained the upheaval, which lasted for more than two months. "They include," Chivers reports, "Western support, the protesters' resolve, cash from wealthy Ukrainians, coaching by foreign activists who had helped topple presidents in Georgia and Serbia, the unexpected independence of the Supreme Court and cheerleading by television station, Channel 5, which Mr. Kuchme never shut down," as well as various elements of Ukraine's security forces.[86] All these forces converged on the Maydan, which itself became an actor in the drama swirling all around it. The Orange Revolution represents a classic instance in which the form and function of the physical urban center can determine a city's history.

The political city joined with the physical city in support of efforts to bring Viktor Yushchenko to power. As noted above, the Kyiv City Council was among the first to reject the results of the November 21 presidential runoff being released by the Central Election Commission. Mayor Omelchenko immediately rejected the use of force to disperse the growing throngs of demonstrators and protesters.[87] The mayor went further, paying back Kuchma for the president's various attempts to drive him from office, by literally keeping the lights on in the Maydan. City officials kept the subways and buses running, the sound stage volume turned up high, and the enormous stadium-sized television screens switched on. City officials spurred on revolution merely by operating as if all were normal.

Omelchenko chose, in the end, to treat the presence of well over a million protesters as if it were just a particularly large turn out for the annual City Day celebrations held each fall. The mayor was rewarded by being reappointed as head of the Kyiv City Regional Administration by President Yushchenko's prime minister, Iulia Tymoshchenko, immediately after the post–Orange Revolution government came to power.[88] Omelchenko nonetheless faces a tough race for reelection in early 2006.

Kyiv shaped the Orange Revolution in another more profound manner: The city infused the protest with a sense of tolerance. The ready acceptance of differences so evident in the Kennan Institute surveys and interviews was on public view every day of the Orange Revolution. The unfolding revolutionary events of November and December 2004 revealed that a democratic middle ground had taken hold within Kyivan life during the decade and a half since Acting Mayor Oleksandr Mosijuk had lifted the blue-and-yellow banner of Ukrainian independence in front of City Hall (on a flagpole readily visible from the Maydan). Kyiv's store of diversity capital had demonstrably expanded.

Many of the eyewitness reports and accounts of the "Maydan Parliament" and its tent city of protesters sound the same theme: The square ex-

uded a peaceful decency combined with a tenacious purposefulness that redefined what it meant to be "Ukrainian." Two particularly perceptive reports
capture the mood on Kyiv's streets.

The Moscow *Kommersant* newspaper correspondent Valeriy Panyushkin
appears to have been overwhelmed by what he saw all around him as he
walked about the Maydan. Panyushkin reported back to his Russian readers an account tinged with pathos and envy:

> Exuberant city. Peaceful, smiling, kind, united people. But most impor-
> tantly—they are free. Free! Free! I experienced jealousy and pride for
> the fact that I am standing among these free and peaceful people. And
> these people were not forcing me out despite the fact that I came from
> Russia, a country whose Minister of Foreign Affairs is low enough to
> make an official statement about NATO's geopolitical claims to Ukraine.
>
> Listen, you, minister, come here, to Kyiv. Go to Maydan and despite any
> orders from the Kremlin, you would not be able to utter a word about
> NATO's geopolitical claims. There are many more of these people—
> young men and women, children and elders—than a Minister or a Pres-
> ident of Russia could ever imagine in their wildest dreams.[89]

The British historian Neal Ascherson caught this same tone on the streets
as well as anyone in a considerably more understated report that appeared
in the *London Review of Books:*

> The demonstrators say that they have discovered a new country, a
> Ukraine they can be proud of. "Before we were a people not a nation,"
> several people said. In this new country, which they are still exploring,
> they will catch up with "other European nations." Once, an older Euro-
> pean nationalism understood this "catching-up" as matching rivals in ar-
> maments, steel production, colonies. Now, what's envied is not missile
> batteries or hydroelectric dams but something called "normality."[90]

Elements of this "normality" include a spirit of compromise, which, in
the end, resolved the immediate conflict over who had won an election. The
"party of power" acceded to new elections, while the "opposition" accepted
a package of reforms to the Ukrainian Constitution that would eventually
devalue the post of president. Transnational migrants were not major participants in the drama at hand (although many evidently participated in the

demonstrations).[91] In their own way, transnational migrants had already helped to nurture the "normality" that sprang so suddenly into view by fostering the creation of a middle ground.

## Being Healthy, Safe, and Wise

Will the presence of so many new Kyivans who do not fit long-held notions of city residence lead to the creation of a new middle ground in an urban community traditionally divided by language? Alternatively, will migrants become one more object of ethnic scorn? Will Kyiv generate new diversity capital? Partial answers to such questions emerge from the Orange Revolution. More complete responses must be based on the reactions of local officials and citizens to specific migrant concerns over health, education, and public security.

Kyiv's transnational migrant community is generally healthy. Consisting primarily of working-age males, the overall migrant population suffers from few of the chronic illnesses associated with childhood and old age.[92] With the exception of the Chinese and Vietnamese communities, Kyiv's migrants are predominantly from the middle class and professional strata of their home countries and, therefore, generally representative of groups with lower health risks.[93] Those migrants who enter the country legally, or who subsequently gain legal status, are subjected to health examinations. Authorities regularly deny entry or residency to those suffering from any number of different infectious and chronic diseases.

Ukrainian health officials maintain that the transnational migrant community is generally more healthy than the national norm.[94] There have been no reports to date of epidemiological issues associated with the migrants' presence in the city and in Ukraine, whereas migrants from Africa and the Middle East in particular show lower rates of HIV/AIDS and tuberculosis infection than the indigenous Kyiv population.[95] Most hospital services provided to migrants are limited to outpatient treatment, with pediatric diseases accounting for the greater portion of migrant admissions. Migrant demand for the city's health care facilities, therefore, remains modest.

Ukrainian law requires that medical assistance at the place of residence be made available free of charge to all the country's officially recognized permanent residents.[96] The reality for native-born city residents and transnational migrants alike is a much more variegated pattern of free, commercial, and "gray" fee-for-service treatment (with the last being paid "under the table"). Physicians' demands for cash frequently become the point at which migrants begin to express concern over the quality of their health care in Ukraine—even migrants who have traveled from war-torn and impoverished communities abroad.

Migrants generally have reported that legally mandated free medical care has been provided in only two-thirds of those instances when it was requested, with one-third of all treatment requiring some form of official or unofficial payment. This overall pattern varies according to migrant group, however, as Pakistanis pay for medical services in 70 percent of their cases of need, Arab and Kurdish migrants report having to pay 50 percent of the time, and Vietnamese migrants pay for more than 40 percent of requested service provision.[97] More important, all migrant communities evaluate the availability of medical services to be more problematic than do native Kyiv residents.[98] Three areas of notable exception are shock trauma treatment for accident victims, medical services relating to childbirth, and pediatric care—all of which receive relatively high marks from migrants.

In other words, Kyiv's transnational migrant communities generally place relatively fewer demands on the city's health care community, and are relatively more dissatisfied with the available medical treatment, than native-born residents of the Ukrainian capital. This pattern of discontent is magnified in those instances when migrants become infected with serious diseases in Ukraine. The problem is particularly troubling with regard to tuberculosis, which has reached epidemic proportions throughout Ukraine. A total of 11 percent of Vietnamese and Pakistani respondents to the Kennan Institute Migrant Survey, 14 percent of Afghan respondents, and more than 30 percent of African emigrants reported having contracted tuberculosis after having arrived in Ukraine.[99]

Although health care in Kyiv falls under the purview of national state institutions rather than local authorities, discontent among all migrant communities over the accessibility and quality of local health care has emerged as a potential major source of tension between non-native-born Kyiv residents and the public authorities. Approximately 70 percent of all respondents to the Kennan Migrant Survey report being dissatisfied with their health care in the city.[100]

Migrant communities appear to evaluate access to local schools more positively than they assess health care. Nevertheless, the growing number of transnational migrant children within Kyiv city schools is beginning to create difficulty for teachers, administrators, parents, and pupils alike. As Nancy Popson has reported, classrooms on Kyiv's outskirts are the prime venue in which native Ukrainian children come into day-to-day contact with new-comers to their society—children from countries like Angola, Afghanistan, Turkey, and Vietnam, and from formerly Soviet countries torn by civil strife.[101] The school represents a microcosm of the larger social environment, where pupils are taught rules of proper interaction and communication with other segments of society at the same time they learn their grammar and multiplication tables. This is critical for migrant children, providing them with a living laboratory in which they can—and must—experiment daily in their quest to master the complicated maze of linguistic and social norms.

In striving to educate children in both academic and social spheres, Kyiv's school system has been stretched by the influx of newcomers. In 2000, it was reported that more than 1,250 children from refugee families seeking refugee status were living in Kyiv. This does not include children whose parents arrived in Kyiv for economic reasons, or who might have ended up in Ukraine illegally.[102]

The children of migrants largely reported using either their native language or Russian when speaking with family and friends in Ukraine. Their parents, however, are choosing disproportionately to send their offspring to Ukrainian-language rather than Russian-language schools (by 58 to 42 percent).[103] This preference is related primarily to parental assessments of the quality of instruction available in Kyiv's Ukrainian-language schools as being of higher quality than in the city's Russian-language schools.[104]

Access to schooling continues to be problematic for some transnational migrant families, primarily for economic reasons. Although primary and secondary education in Ukraine technically is free of charge, the expenses related to school (uniforms, books, "voluntary" donations for school renovations) mount up quickly.[105] Parents may decide not to send their children to school due to a lack of language skills or because they do not plan to stay in Kyiv for the long term.

Another reason parents withhold their children from school appears to be that they erroneously believe they will be required to provide official residency permits for their children to attend Kyiv class. Under Ukrainian law, schools must accept children whose parents can provide documentation of the child's full name and birth date. In practice, it is unclear to what extent such regulations are followed. Popson, for example, has observed that three of four school directors from schools with large migrant enrollments understood and followed these regulations.[106]

As is the case in many cities with large migrant populations, migrant families are concentrated in a few neighborhoods, thereby permitting national and city officials to ignore the changes that are occurring in the micro-demographic school environment. Those schools in Kyiv with the largest numbers of migrant children are found in the districts surrounding large markets or bazaars, such as Troieshchina.

Given the current fiscal and administrative crisis in Ukraine, teachers and directors at many schools with large migrant student enrollments are being left without significant resources—human or financial—to deal with issues that arise due to the influx of the new children. At the level of the national Ministry of Education, the ministry's Department for the Defense of Children's Human Rights is charged with most of the questions pertaining to migrant children. However, that department is very small and focuses more on rights of access than on issues of second-language education and cultural pluralism in schools. Moreover, ministry officials, like their counterparts in City Hall, have been slow to recognize and adopt to the new demographic realities of the Kyiv classroom.[107]

Reports of the inability of the local school system to adapt to the presence of transnational migrant children in their classrooms are offset by the generally high level of academic achievement among migrant students. With the exception of African children—whose exclusion from the general pattern once again appears to be a consequence of rampant racism in Kyiv—teachers, administrators, and parents alike report that the academic performance of migrant children is comparable to that of native-born Kyivan children.[108] Vietnamese, Chinese, Arab, Kurdish, and Pakistani parents have become actively involved with their local schools—at least at the level of meeting with and talking to their children's teachers.[109] School attendance rates among African, Vietnamese, Chinese, Arab, and Kurdish children equal or surpass those of indigenous, native-born Kyiv students.[110]

Without the support of national and municipal institutions, efforts to better accommodate migrant children must be carried out at the individual school level. Such programs will be difficult to sustain without sufficient state funds and experienced faculty. Teachers barely getting by on their measly wages are often less than enthusiastic about having to volunteer for extra classes for which preparation would likely be more intense.[111] Consequently, there has been growing tension between migrant communities and Kyiv authorities over the formal arrangement of their children's education.

In addition to concern over the quality of local health care and educational services, Kyiv's migrant communities hold all branches of the Ukrainian police in considerable disrepute. Two-thirds of respondents to the Kennan Institute survey—including more than three-quarters of the Afghans and Africans participating in the survey, and more than half the survey participants from the Middle East and Pakistan—reported having heard of injustices committed against migrants in Kyiv.[112] By far the greatest number of complaints of unjust actions were lodged against the police—ranging from a low of 60 percent of the Pakistani respondents who had heard of injustices being committed against migrants to a high of more than 90 percent of similarly responding Vietnamese participants.[113] Between one-half and two-thirds of respondents from all migrant groups report having personally been the victim of an unjust act by the divisions of the Ukrainian national police responsible for order in Kyiv.[114] Interestingly, complaints about police injustices are highest among those respondents who are in the country legally. By contrast, just under 40 percent of all respondents in the Kennan Migrant Survey sample registered similar complaints of injustice against the city's criminal gangs.

Migrants, of course, were not alone in their distrust of the Ukrainian police (at least before the Orange Revolution). When asked what problems are likely to be faced by transnational migrants to their city, a majority of native-born Kyivans identified police officials as the leading nemesis for their migrant neighbors.[115] Such complaints are based on personal experience and observation. When queried about their level of trust of the police, 61 percent of native-born respondents to the Kennan Migration Project survey of the general population expressed some level of distrust. Moreover, when asked to evaluate the relations between the police and city residents along a 5-point scale (with a high of 5 and a low of 1), the average rating for citizens of Ukraine and for migrants alike was nearly identical—2.2 points for the first group as opposed to a rating of 2.0 points for the migrants.[116]

The difficult relationship between migrants and police authorities represents a third potential flash point for future community unrest. The issue is a knotty one, however, because beat-level police justify their actions by what they proclaim as a higher level of criminality among migrant communities. The fact that the reverse is largely true—Kyiv's migrant communities are generally less prone to criminal activity than the city's native-born population—matters little when a migrant is stopped on the street in a document check.[117]

The suspicion with which some police authorities in Kyiv view transnational migrants points to a more difficult problem inhibiting assimilation into Ukrainian life. The city's sensationalist media—as well as politicians hungry for attention—increasingly has waged slander campaigns against the city's newest arrivals.[118] Concern over the supposedly high incidence of criminal activity associated with migrants figures prominently in such attacks, despite the fact that the reality is quite different.[119] The impact of such accusations undercuts the ability of migrants to carve out a place for themselves in Kyivan life. False media coverage of migrant issues represents a highly pernicious trend in a city that is becoming the recipient of significant migration from abroad.

## Media Ignorance and Backlash

One would never know from the city's media that Kyiv's transnational migrant communities are healthier and less prone to crime than its indigenous population. Moreover, local newspaper readers and television viewers probably would be surprised to learn that the children of transnational migrants are performing as well in school—or as poorly—as their native-born classmates.

The issue is not a lack of coverage of the presence of migrants in the city or country. Surveys of local and national press sources—including particularly comprehensive reviews conducted for the International Organization for Migration by the Kyiv sociologist Mykola Shul'ha, as well as others carried out over the years of Ukraine's independence—have shown that national publications devote a comparatively large amount of space to questions of migration.[120] Rather, the problem is a distinctly negative portrayal of migrant communities, a depiction that often distorts a far more positive migrant reality on city streets. More important for this study, such an inaccurate clamor of concern undermines the city's capacity to accommodate diversity.

Shul'ha's systematic review of national and regional Ukrainian newspaper coverage of migration issues between 1991 and 1999 underscores the extent to which the Ukrainian public and migrants are being poorly served by the local Fourth Estate. Nearly half the articles reviewed by Shul'ha and his colleagues from ten major newspapers focused on crime in some form. The press wrote more frequently about illegal immigrants than about those who had come to the country legally. Coverage included melodramatic stories about smuggling rings that bring migrants—and illegal weapons and narcotics—in and out of Ukraine. Migrants themselves are shown as frequent perpetrators of criminal acts and hooliganism as well.[121]

In addition to criminality, Shul'ha observes that migrants are accused of bringing disease into the country. The Ukrainian press, according to his survey, has been quick to blame migrants for outbreaks of malaria, tuberculosis, cholera, and HIV/AIDS.[122] Such tracts, as already noted, ignore evidence to the contrary presented by Ukrainian health professionals.

Shul'ha is concerned not only with the number of articles appearing in Ukrainian newspapers that present negative images of transnational migrants but also with the vehemence of their portrayal. Just as important, the press has presented few positive images of transnational migrants, rarely raising the complex economic, social, cultural, and humanitarian issues surrounding the arrival of migrants to Ukraine. Media outlets hardly ever mention the wars from which many migrants are fleeing when they arrive in Ukraine.[123]

The trend over the years since Shul'ha's work has been, if anything, even less responsible and more disturbing.[124] Some articles have moved beyond the mere "reporting" of criminality among migrants to suggest the possibility of violence against migrants by angered Ukrainians.[125] Some authors writing in otherwise respectable journals in recent years have been moved to speak favorably about death threats against migrants who are somehow destroying the purity and calm of Ukrainian life.

Media reporting reflects an increasingly destructive public dialogue on the migration issue. Shul'ha argued that media accounts reveal a disturbingly primitive conceptual framework for the discussion of migration issues in Ukraine. News reports reinforced an inability among Ukrainian elites to engage in considered public analysis of the problem. Never having faced the issue before independence, Ukrainian opinion leaders often lack the conceptual tools to think through the issues raised by the presence of transnational migrants in their city. Consequently, depictions of migrant life in Ukraine quickly descend into muckraking sensationalism.[126]

Negative and misleading press coverage of migrant issues has been especially disturbing because the media often shapes perceptions and modes of discourse that resurface in more overtly political contexts. Karim Karin has argued, for example, that negative press coverage of multiculturalism in Canada during the 1980s eventually found an outlet in the ever-more bellicose political discussion of issues relating to multiculturalism—including immigration—by the late 1990s.[127]

Evidence of mutually reinforcing influences between antagonistic press reports and antimigrant political rhetoric is beginning to grow. If, in 2000, a well-placed former Kyiv official could declare that there was not, as yet, "a Ukrainian Le Pen," the stage was being set for demagogic antimigrant populism just a few years later.[128] A group of Kyiv candidates in the 2002 parliamentary elections formed the Kyivska Fortetsia (Fortress Kyiv) electoral alliance and distributed campaign literature linking Muslims to international terrorism.[129] Local Kyiv politicians—at times on the urging of American campaign consultants—are learning how to use appeals to people's fear of difference as a means for mobilizing voters around their candidacies. Their success depends to a considerable extent on previous press reports that misrepresented the place of transnational migrants in local life.

Kyiv increasingly finds itself with a sizable population of residents who do not fit neatly into long-held conceptions of the city as a center of Russian and Ukrainian—and, to a lesser extent, Jewish—life. Popular discourse about the city has centered on competing images of Kyiv as either the "Mother of Russian Cities" or as "Capital of Ukraine." Politicians nurtured in a Soviet system that excluded the possibility of meaningfully discussing social policies have proven themselves ill prepared to meet the challenges posed by the arrival of large numbers of new residents from abroad. Police officials have revealed their predatory nature, pouncing on migrants as another source of tribute to be demanded from local residents. Journalists have failed to inform their readers, listeners, and viewers of the complex realities of transnational migrant life in Kyiv.

"Official" Kyiv—as opposed to broad urban society—appears to be providing little space for transnational migrants to create a new "middle ground" in their city, at least before the Orange Revolution. Admittedly, a postrevolution "normal" and "democratic" Ukraine could well respond more effectively to the needs of transnational migrants.

Beyond local elites, the atmosphere of tolerance so visible on Kyiv's streets during November and December 2004 strongly suggests that the city's diversity capital had, in fact, increased over the course of the previous decade. Even more time will be required before it will be possible to determine just how that enhanced acceptance of differences will (or will not) become translated into increasing "urban social sustainability." The core of diversity capital, "urban social sustainability," as defined by Richard Stren and Mario Polese, is the existence of "policies and institutions that have the overall effect of integrating diverse groups and cultural practices in a just and equitable fashion."[130]

For their part present, migrants have remained strikingly optimistic as they have evaluated their lives in the Ukrainian capital.[131] They have embraced the country's upheavals just as they welcomed Ukraine's liberal immigration and citizenship laws.[132] They consistently speak of tolerance in the attitudes of the Kyivans whom they meet on city streets every day; and they speak of respect for the schoolteachers and health professionals whom they encounter in trying to live out their lives. For migrants, a "middle ground" would seem to exist already, the more so as they have looked to their immediate neighbors and coworkers rather than to officials and commentators.

## Civic Tolerance in the Wake of Revolution

Ukrainians in general, and Kyivans in particular, have expressed a relatively high degree of tolerance in a number of social science studies conducted before they took to the streets in November 2004.[133] Kyiv residents have revealed themselves in polls and in behavior to be generally accepting of the presence of various ethnic and religious groups in their city. Though attitudes remain considerably more inchoate with regard to migrants from outside the old territories of the Soviet Union, acceptance levels for all groups rise as education increases.

These patterns were sustained throughout the 1990s despite a broad increase in intolerance that accompanied the economic decline of the immediate postindependence years. The first reaction of the typical Kyivan to someone different from himself or herself appears to be quiet forbearance.

Acceptance levels decline when city residents are questioned about more narrowly defined relationships—such as having members of a specific ethnic, confessional, or linguistic group as neighbors, coworkers, classmates, family members, and the like.[134] Broad proclivities toward tolerance are not sustained across racial groups. Initial post-Soviet surveys indicated considerably more distrust and hostility toward visible minorities (Vietnamese, "Arabians," "Blacks," and "Gypsies") than among groups of European heritage.[135] Moreover, a majority of respondents to the Kennan Institute survey of Kyiv residents stated that they would not like to have immigrants from Asia and Africa as neighbors, close friends, or family members.[136] These findings are consistent with the above-mentioned feelings of enmity perceived by African respondents to the Kennan Institute Migrant Survey.

Nevertheless, the more remarkable story has been the emergence of a wide reservoir of tolerance even while acknowledging the existence of latent hostility toward specific migrant communities. A majority of Kyiv residents are sympathetic to the challenges faced by migrants in their city.[137] This empathy is especially noticeable among Kyivans of all backgrounds who had been in the city for fewer than five years.[138]

Generally accepted standards of public decorum further reinforce an atmosphere of tolerance—as was seen during the Orange Revolution. Even those Kyivans who harbor negative attitudes toward migrants deem polite behavior to be the norm to be followed when interacting with others who are different from themselves. The result has been a general sharing of public space among all groups in the city even when meaningful contact among various groups remains slight.

This pattern of public acceptance appears to be linked to the city's long history of ethnic, linguistic, and confessional cohabitation, which has been reinforced by the Ukrainian state's efforts to nurture a civic identity that transcends group boundaries.[139] Such policies are a consequence of the stark internal divisions among ethnic Ukrainians and Russians, Ukrainian-speakers and Russian-speakers, and other groups traditionally associated with the city (e.g., Jews and Greeks). Kyiv's harsh and bloody twentieth-century history may have played a role as well—having demonstrated the shamefully violent consequences of intolerance in daily life.[140]

The large proportion of the city's population who themselves were born outside of Kyiv—be they migrants from elsewhere in the Soviet Union for from within Ukraine—similarly emerges as a significant contributing factor to expressions of sympathy for and tolerance of transnational migrants in the Ukrainian capital.[141] Those who moved to the city are distinctly more tolerant of other migrants to the city than native-born Kyivans. Tolerance toward transnational migrants becomes strongly linked to feelings of empathy toward those who have lost their homes and are vulnerable newcomers to the city's life. The hostility that was reported in surveys—with the notable exception of racist attitudes toward African migrants—appears to have been based on the absence of information rather than any inherent malevolence.[142]

The gentle atmosphere of acceptance reported by transnational migrants themselves becomes more explicable when viewed through the prism of survey data exploring popular attitudes. Sufficient social space and diversity capital already exist in Kyiv for transnational migrant communities to occupy a "middle ground" between local Ukrainian and Russian linguistic and ethnic communities. The conceptual framework found among city residents (as opposed to many city officials) has space within it for accepting people of difference within the city. This broad tolerance has been visible both in local responses to migrant communities and in the behavior of Kyiv residents during the events now known as the Orange Revolution.

## Creating or Destroying Diversity Capital: Still an Open Choice

Popular tolerance stands at odds with what is often reported in surveys of attitudes toward migrants in other post-Soviet states. More significantly, it contrasts with the increasingly aggressive language used by the local media (and by some politicians) when discussing the place of transnational migrants in Kyiv. This reserve of positive feeling can advance the city as a supportive place of residence for many transnational migrant communities. The experience of one Afghan migrant suggests that Kyiv may still come to fit into the increasingly virtuous patterns identified in Montreal and Washington.[143]

Mahmoud and his family fled Afghanistan in 1993 while he was in his early thirties. They arrived in Ukraine after having traveled for most of that year through Pakistan and Kazakhstan. Intending to find his way into Europe, Mahmoud never expected to remain in Kyiv for nearly a decade. Mahmoud, his wife, and their daughter eventually received official refugee status in Ukraine after paying $450 in bribes in 1998. His family secured an apartment near the robust Troeshchyna market, added a second daughter, and began to build a life in Kyiv. By 2001, he had invested in an iron door and window-bars on his apartment to protect the world that he was building for himself from thieves and bandits.

Mahmoud and his family have rarely received help from Ukrainian authorities and only marginal assistance from international refugee organizations. They have made their way in Ukraine on their own. Both daughters attend Ukrainian-language schools, with Mahmoud and his wife periodically participating in various parent–teacher meetings. He is well connected to Afghan community organizations and prays daily at a local mosque. He can communicate in spoken Russian and Ukrainian, though he claims not to be able to write in either language. A diabetic with a crippled right arm, he has received appropriate medical attention in Kyiv through a variety of channels. His earnings from trading at the Troeshchyna market provide a living for his family in excess of what he could have sustained either in Afghanistan or Kazakhstan.

Mahmoud personifies everything that all too many politicians and journalists have refused to see when they look at and speak about Kyiv's new migrants: a quiet tale of success. He and his family have navigated the difficult and complex world of Russian-Ukrainian relations in Kyiv, creating a life for themselves unthinkable elsewhere.

The larger significance of the travails and accomplishments of Mahmoud and his family may well be that Kyiv and Ukraine are becoming home to thousands of other international migrants with similar histories to tell. So many "Mahmouds" had moved to Ukraine by 2003, in fact, that the number of requests to obtain Ukrainian citizenship that year exceeded the number of native-born Ukrainians officially moving out of the country for the first time since independence.[144] By their quiet achievements, Mahmoud, his family, and thousands like them are establishing a middle ground in a country and a city long bifurcated by divisions between Russian and Ukrainian linguistic and cultural communities. His story provides evidence that, on balance, contemporary Kyiv has created more diversity capital than it has destroyed.

# Part IV

# Concluding Observations

# Chapter 7

# Diversity Capital Created

The picture of the city that we carry in our mind is always slightly out of date.

—Jorge Luis Borges, "Unworthy," 1998[1]

If Montreal has become an intercultural city rooted in a French-language milieu, and Washington has demonstrated a potential to reinvent itself as an intercultural city rooted in an African-American milieu, the future of an intercultural Kyiv remains elusive. Thousands of new residents have arrived from beyond the boundaries of Ukraine, frequently from the Middle East, Asia, and Africa. The city's potential as an intercultural metropolis embracing dozens of migrant communities has become apparent. Kyiv's indigenous population has demonstrated a capacity to respect migrants from a variety of ethnic, religious, and linguistic backgrounds. Whether or not municipal institutions and the media can adapt to this new reality in a manner that advances social harmony is unknown, even as the success of the Orange Revolution suggests the sensibility of optimism.

Despite these uncertainties, Montreal, Washington, and Kyiv are different cities than they were a few decades ago. They are more comfortable with diversity; they have a greater capacity to accommodate diversity—both a formal capacity in official institutions and an informal capacity in changing patterns of social relations; they have accumulated greater stores of diversity capital. Montreal, Washington, and Kyiv are closer to resembling great cosmopolitan entrepôts such as New York and Amsterdam in 2005 than they were in 1965.

Their very real accomplishments have not been easily secured. These achievements have been made possible by the arrival of thousands of new residents in recent years who are quite unlike previous Montrealers, Washingtonians, and Kyivans. The deep linguistic, racial, and confessional divisions that marked all three cities withstood decades of dramatic economic restructuring and political reorientation, the ever-present pressures of globalization, and massive urban sprawl that began in earnest a half-century ago. Only the presence of increasingly diverse groups and cultures could break down long-standing patterns of elite, racial, and linguistic accommodation.

As in New York and Amsterdam of the past and present—and as in so many other commercial centers—Montreal, Washington, and Kyiv are increasingly fragmented urban communities in search of renewed common ground now that no single group can attain permanent dominance.[2] To achieve any set of personal or group goals, individuals and communities must transcend a "zero-sum game" by engaging with others to pursue shared objectives. Twenty-first-century Montrealers, Washingtonians, and Kyivans can no longer afford the practical inconveniences of division quite as readily as in the past. The very meaning of such stark and separate realms has been called into question by the growing presence of tens of thousands of residents who fit in neither camp. Citizens and elites alike have increasingly little choice but to practice the art of pragmatic pluralism, even as their parents and grandparents once stared at one another across tightly defined spatial, racial, ethnic, and linguistic divides.[3]

In other words, Montreal, Washington, and Kyiv have considerably greater capacity and motivation to accommodate diversity than a generation or two ago. All three cities, together with their metropolitan regions, have enhanced their capacities for "urban social sustainability," which Richard Stren and Mario Polese identify as "policies and institutions that have the overall effect of integrating diverse groups and cultural practices in a just and equitable fashion."[4] Francophone Montrealers still tangle with their anglophone neighbors; black Washingtonians still have their battles with their white neighbors; and Ukrainian Kyivans still confront Russian Kyivans over language. Yet all three of these long-divided urban communities exhibit considerably more diversity capital than was the case just a few decades ago. This change is a direct consequence of the arrival of groups of migrants who do not fit neatly into previous psychological, social, cultural, and legal categories.

The recent experiences of Montreal, Washington, and Kyiv thus underscore an observation made by Myron Weiner a decade ago: "Solutions to the problems posed by unwanted transnational migration cannot easily be formulated into sound bites."[5] All three cities' experiences with negotiating the presence of large transnational communities are sufficiently similar for comparison. William Rogers Brubaker's admonition that migration forces societies to reshape their institutions—and to rethink the meaning of citizenship—is of direct relevance to the burgeoning transnational migrant communities in Montreal, Washington, and Kyiv alike.[6]

## Creating New Gateways

Nancy Foner demonstrates the power of such research in two books: *New Immigrants in New York* and *From Ellis Island to JFK: New York's Two Great Waves of Immigration.*[7] Foner's focus on New York City is hardly accidental. As Larry S. Bourne observes in relation to another great migrant metropolis, Toronto (a city in which more than 40 percent of the metropolitan population is foreign born):

> Cities, especially large metropolitan areas, sit at the intersection of transitions in the economy, demography, and social order, as well as in culture, technology, and politics. They are, in effect, the local venues where most innovations occur and where the impacts of external forces are most prominently expressed. They are the arenas in which economic linkages and social networks are constructed and deconstructed, and where political conflicts that invariably occur within and between these forces of change are worked out.[8]

More specifically, Bourne continues, "The most obvious example of globalization, in the sense of being tied to the rest of the world, is foreign immigration."

For Foner, cities become the locus of migration chains and economic networks in which brokers move easily among minority communities and societies at large. Those brokers—ranging from street market vendors to international bankers—who integrate migrant communities into the host society are most often concentrated in just a few gateway cities.[9] Urban life transforms migrant groups into ethnic communities with shared memories and perceptions, for it is on city streets that migrants come to appreciate their own similarities in opposition to the world around them.

Unlike Foner's New York—or Bourne's Toronto—Washington and Kyiv have not always been gateways into their own societies. Montreal arguably entered a new phase in its development as a gateway city nearly a half-century ago, when Canadian migration policies regarding Africa, Asia, the Caribbean, and Latin America underwent profound change. As a result, all three cities represent something quite different from New York and Toronto, namely, communities long divided in two by such factors as language and race.

Montreal, Washington, and Kyiv reveal a different tale from that represented by New York or Toronto (or Amsterdam). Their histories are stories of relentless struggle by the vanquished—French, African-American, or Ukrainian—to persevere in the face of both open hostility and mere indifference on the part of the victor—English, European-American, or Russian. Conqueror and conquered long shared an interest in sharply defined boundaries: the English, whites, and Russians as a means of asserting superiority; the French, blacks, and Ukrainians as a means of survival. Mutually reinforced borders defined local life, whether literal streets—Saint Lawrence Boulevard / Boulevard Saint-Laurent in Montreal and Sixteenth Street, Northwest, in Washington—or the cultural, confessional, psychological, and linguistic divisions in Kyiv. Scant impulse emanated from within these quite different Canadian, American, and Ukrainian cities to bring about a commitment to a shared middle ground among all city residents.

Much has transpired during the course of the past half-century to change these realities. The process of transformation began slowly, remaining obscured at first by communal violence in Washington (e.g., the 1968 riots) as well as in Montreal (e.g., the 1970 "October Crisis"), and muffled under a blanket of Brezhnev-era stagnation in Kyiv. A new understanding of Kyiv as a center of democratic values became fully visible only with the onset of massive street demonstrations protesting fraud in the 2004 race to elect a new Ukrainian president.

As just noted, neither Washington nor Kyiv was a major magnet for transnational migrants before the 1970s. Though Montreal has a longer and more fruitful experience of assimilating transnational migrants, the color and composition of that city's immigrant communities underwent substantial change around the same time. New transnational migrants to all three cities have come to occupy a middle ground between entrenched linguistic and racial communities; quite literally so in Montreal and Washington. The stark boundaries represented by Saint Lawrence Boulevard / Boulevard Saint-Laurent and Sixteenth Street, Northwest, have begun to blur as Asian and Central American migrants have moved into such neighborhoods as Côtes-des Neiges and Nôtre-Dame de Grâce, Adams Morgan and Mount Pleasant—and Troeshchyna.

The workplace and the classroom have become at times uneasy meeting grounds for a variety of cultures, religions, and linguistic groups, as well as racial and ethnic communities. Perplexingly complex and finely tuned linguistic hierarchies have appeared in the Montreal labor market at the same time as children of European heritage have become a minority presence in school districts throughout the Washington metropolitan region. All three cities are becoming gateways through which transnational migrants pass into a larger host society.

Conflicts over language and race continue to dominate local politics, to be sure. Aggressive language laws coerced Montreal's transnational migrants to send their children to French-language schools; migrants to Washington from countries such as Brazil suddenly discovered that they are, indeed, "black" as well as Latin American. Bitter elections of all sorts were fought over questions of language and race. By the 1990s, the outcome of contests at the polls no longer seemed quite predictable. Sovereignty for Quebec went down to the narrowest of defeats in a brutal October 1995 referendum; white politicians controlled a majority of the seats on the D.C. Council by the end of the 1998 electoral season; and pro-Western reformers overwhelmingly carried Kyiv to help bring Viktor Yushchenko to office in late 2004.

Neither language nor race have ceased to define realities in Montreal, Washington, and Kyiv. To borrow once more from Elke Laur, their "daily dance of life" continues to be mediated through language and race.[10] Nonetheless, other filters are coming to shape Montreal and Washington. A new pragmatism may be seen in the sort of politicians who stepped to the fore in all three cities by the late 1990s. Dull questions of service provision and metropolitan amalgamation have come to dominate local political discussions. The nature and place of transnational migrant communities have become ever more critical to political, economic, and cultural life. The sort of middle ground that Thomas Bender argues has always been present in New York is taking shape in Montreal, Washington, and Kyiv as they are converting themselves into gateway metropolitan regions.

The impact of expanding transnational communities on local politics and life is real even when the migrants themselves may remain in the background. Migrants played only marginal roles in the battles over demerger on Montreal Island, baseball stadiums in Washington, and constitutional provisions in Ukraine. The outcomes of those conflicts nonetheless reflected deep changes that had taken place within Montreal, Washington, and Kyiv more generally, as Tara Hecksher, Kelly and Maze Tesfaye, and Mahmoud went about building their new lives.

The transformation of neighborhoods, workplaces, and schools prompted by the arrival of tens of thousands of transnational migrants had created a new political middle ground. Former Montreal mayor Pierre Bourque could well not have endorsed interculturalism, former Washington mayor Marion Barry could well not have found himself at press conferences surrounded by white D.C. Council members similarly outraged by proposals from Major League Baseball, and Kyiv mayor Omelchenko could well not have cranked up the volume on the city-controlled Orange Revolution outdoor concert stages had not Montreal, Washington, and Kyiv become different cities. Tara, Kelly, Maze, and Mahmoud are as responsible for their new cities as Bourque, Barry, and Omelchenko. They are present even when they do not appear on the local news coverage of the latest political battle of the day.

## Global Suburbs

The arrival of new, ever more differentiated communities of transnational migrants accompanied two equally profound—almost seismic—shifts in Montreal and Washington life. First, both cities have been transformed from center-city-dominated urban communities into massive urbanized super-regions spreading out along highways and suburban cul-de-sacs for miles in every direction. Second, both Montreal and Washington have carved out distinctive niches for themselves in an ever-globalizing economy.

Mass suburbanization has forever changed the meaning of "city" around the world. The rapid development and easy access to housing, jobs, and services on the periphery of historic cities has converted center-city Montreal and downtown Washington into merely one more node in multinucleated, sprawling metropolitan networks. Residents of both metropolitan regions have far greater choice over where to live, work, and go to school than has ever been the case. Traditional spatial communities dissolve in the face of development pressures. Quebec's wealthiest community is no longer anglophone Westmount but a number of francophone suburbs along the Island of Montreal's north shore; the District of Columbia is now one of the least populated component jurisdictions of its own metropolitan region. Older boundaries between French and English, and between black and white, are submerged beneath flyovers, cloverleafs, and ubiquitous shopping malls. This new North American spatial reality reinforces a middle ground being created by large communities of migrants from abroad.

Suburbanization is driven by the fresh economic realities whereby both Montreal and Washington are redefining their places in the global economy. Montreal assertively markets itself as the only French metropolis in the Americas—attracting transnational companies and associations desirous of access to both French-speaking and American continental markets and publics. Washington is no longer merely a government town but also a dynamic leader in various international informational and biological high-technology industries. Such employers, arriving concurrently with suburbanization and the new transnational migration, are far less rooted in traditional definitions of place. Their offices, and those who fill them, are distributed across vast metropolitan regions.

Montreal's and Washington's new metropolitan spatial and economic realities have meant that transnational migrants are not relegated to a single—low-paying—segment of the local economy. Moreover, the presence of migrants means that their suburbs no longer merely recreate the linguistic and racial divisions of the past. Foreign-born Montrealers and Washingtonians may be taxicab drivers and computer specialists, construction workers and medical researchers, restaurant waiters and university professors. Transnational migrants occupy various rungs on the entire Montreal and Washington income hierarchies. Some discrete segments of migrant communities in both Montreal and Washington earn, on average, more than segments of native-born poor people.

## No Longer What They Once Were

Put simply, neither Montreal nor Washington is the city it was just half a century ago. One can no longer speak of secure linguistic and racial hierarchies. Both metropolitan regions have become redivided, dispersed, and recontested by an endless array of social, economic, confessional, linguistic, ethnic, and racial boundaries. Notions of community and virtue are no longer shared; compliance must be earned rather than asserted. Transnational migrants are forcing their native-born neighbors to renegotiate all types of social, economic, and political arrangements.

Both Montreal and Washington survive and grow on the basis of pragmatic decisionmaking, which creates the space for a pluralism of personal and group identities and goals. Montreal's "two solitudes" have become many. The Island of Montreal constitutes an intercultural metropolis rooted in a French-language milieu. Though one cannot yet speak of Washington as representing a cultural diversity rooted in an African-American milieu, such a point of view becomes ever more appropriate as the city of Washington and its metropolitan region redefine long-held conceptions of community, place, and space. This possibility of an intercultural Montreal secured by a French-language core—and of an intercultural Washington embedded in an African-American foundation—was simply unthinkable when army troops occupied both cities just a little over a generation or two ago.

Admittedly, Kyiv's place in this discussion may be premature. The Ukrainian city serves as the capital of a state too newly born to have either well-formed conceptual or institutional definitions of its purpose and place in the world. Kyiv thankfully has avoided the sort of communal violence that marked a turning point in Montreal's and Washington's histories. In August 1991, Mayor Grigorii Malashevs'kii managed to make it back to the city from a Crimean beach in time to negotiate keeping Red Army tanks at bay. In November 2004, Mayor Oleksandr Omelchenko ordered local security forces not to resist the swelling Orange Revolution crowds.

Surveys of Kyiv's residents—and of transnational migrants now living in the city—reveal a labyrinthine urban reality in which a facility toward diversity exists at the societal level even as difference is denied by all too many local elites. Large migrant communities have come to live in several Kyiv neighborhoods, create jobs for themselves and for native-born Kyivans alike, educate their children, and fall ill throughout the Ukrainian capital. Conceptions of a Kyiv predicated on Ukrainian and Russian linguistic competition no longer encompass local life. Thousands of Afghan, Kurdish, Chinese, Vietnamese, and African Kyiv residents challenged the legacies of the Soviet and pre-Soviet past by their very presence. By doing so, they contributed in small yet meaningful ways to the formation of a new conception of what the city was about—a democratic conception predicated on Ukraine's integration into the world at large.

As in Montreal and Washington, the arrival of new transnational migrants has coincided with a dramatic physical expansion of the city into a sprawling metropolitan region (though Kyiv's suburbs tend to be in the form of prefabricated cement high rises rather row upon row of "quaint" single-family houses). Migrants' presence in the city signals the uneven but real integration of the Ukrainian economy into global networks. As in Montreal and Washington, the dramatically variegated success of individual migrants as well as of migrant communities scatters Afghan, Kurdish, Chinese, and African Kyivans throughout the city's economic hierarchy.

Numerous transnational migrants are better educated, healthier, and more successful economically than many native-born Kyiv residents.

If Montreal has become a metropolitan region of multiple communities sharing an urban life rooted in the French language—and Washington appears to be evolving toward a tangled regional assemblage both rooted in, and simultaneously transcending, the city's difficult racial history—Kyiv remains open to an intercultural future based on the use of the Ukrainian language in official transactions. Taken together, all three cities epitomize different manifestations of a possibly shared twenty-first-century destiny.

## Break Open the Cocoon

Montreal, Washington, and Kyiv are hardly unique among twenty-first-century cities. Similar observations may be made about any number of metropolitan regions around the world that have long been tormented by bifurcated conceptions of place. Transnational migrants have transformed—and are changing—Japanese-Korean relations in Osaka, French-Flemish relations in Brussels, and Catalan-Castellón relations in Barcelona.

In Osaka City, for example, 122,063 "foreigners" were officially registered in 2003 in a city of 2.6 million residents.[11] The city's historic Korean community accounted for three-quarters of this group, having been classified as "foreign" even though many Osakans of Korean heritage may be second-, third-, and even fourth-generation residents of Japan. The remaining quarter includes Chinese, Filipino, Brazilian, Thai, and Peruvian transnational migrants who have few previous ties either to Osaka or to its local native Japanese and officially "foreign" Korean residents.

Beyond traditionally divided cities, a number of societies and cultures around the world appear to increasingly rely on transnational migrant labor even though they remain uncomfortable with diversity. In 2000, the Population Division of the U.N. Department of Social and Economic Affairs calculated the average annual net number of migrants from abroad that would be required to arrive in various countries each year during the next fifty years to maintain a working-age population of similar size to that in 2000.[12] In other words, the United Nations attempted to identify those societies that would require the highest number of transnational migrants to sustain current their labor force capacity. Italy and Germany topped the list, followed by Japan and Russia in a virtual tie for third place. All four of these countries once exported migrants rather than received them. Now cities in all four confront the economic necessity of changing long-held cultural attitudes about transnational migrants, as well as any number of noisome legal, political, and institutional arrangements. Urban communities throughout Italy, German, Japan, and Russia increasingly face the challenge of enhancing their capacities for "urban social sustainability," of expanding their reserves of diversity capital.

Montreal, Washington, and Kyiv have been presented here to illustrate how such broad processes modify urban reality every day as capital and human beings circulate ever more rapidly on a global scale. Their stories reveal the power of cities to absorb new life, to transform themselves, and to provide new opportunities for long-term and short-term residents alike. They represent examples of how a city's diversity capital might expand. And in doing so, they demand attention at a moment in history when cities around the world must learn how to enhance their own social sustainability.

Heightened anxiety over terrorism has cast more suspicion on both the city as a social form and transnational migration as a social phenomenon. The impulse to draw oneself into a cocoon of homogeneity increasingly undermines the celebration of difference. The recent experiences of Montreal, Washington, and Kyiv highlight an alternative strategy for confronting the uncertainties of a dangerous world. All three cities represent lively alternatives to a twenty-first-century metropolitan future in which everyone seeks the false protection of fortress communities. Despite all the imperfections and tragedies evident in these three cities, their robust intercultural vitality suggests possible sources for successful strategies to create and accumulate diversity capital.

# Appendix: Research Design and Methodology for the Kyiv Surveys

Why do people move? What makes them uproot and leave everything they've known for a great unknown beyond the horizon? Why climb this Mount Everest of formalities that makes you feel like a beggar? Why enter this jungle of foreignness where everything is new, strange, and difficult? The answer is the same the world over: people move in the hope of a better life.

—Yann Martel, *The Life of Pi* (Edinburgh: Canongate, 2001), 77

The three surveys conducted in Kyiv in 2001–2 (of transnational migrant families, indigenous Kyiv residents, and Ukrainian specialists and officials working with migrants) were supervised by a binational research team consisting of Olena Braichevska (Kyiv Slavic University), Olena Malynovska (Academy of State Government of the Office of the President of Ukraine), Nancy E. Popson (Kennan Institute, Washington), Yaroslav Pylynskyj (Kennan Kyiv Project), Blair A. Ruble (Kennan Institute, Washington), and Halyna Volosiuk (Ministry of Labor and Social Policy of Ukraine). The researchers Volosiuk, Malynovska, and Braichevska previously worked at the Ukrainian government's State Committee for Nationalities and Migration. Their research was carried out with the support of the George F. Kennan Fund of the Woodrow Wilson International Center for Scholars' Kennan Institute, and with the assistance of the U.S.–Ukraine Foundation as well as the Office of the United Nations High Commissioner for Refugees (UNHCR) in Ukraine.

As just noted, the overall study consisted of three interrelated parts: a written questionnaire for immigrants, a survey of Kyiv city residents, and an experts' survey of officials from various government levels and special-

ists who have regular contact with—or work directly with—transnational
migrant communities. Survey questions focused on the main characteristics
of Kyiv's transnational migrants, their living conditions, interethnic relations, and the attitude of Kyiv city residents and the municipal government
toward them.

## Stage One

The first stage of research consisted of interviews with 233 immigrant
households carried out between June and December 2001. The survey team
developed a questionnaire consisting of twenty-four sections, with an average of ten questions each. The sample included families of immigrants from
African and Asian countries who were living in the city of Kyiv at the time
of the survey. Immigrants from countries that were once within the borders
of the Soviet Union were not included.

Reliable statistics for the total numbers of potential interviewees, and
their countries of origin, were difficult to obtain due to the vague legal status of many in the transnational migrant community (some migrants did not
have national documents, permission to stay in Ukraine, permits to work or
conduct business, etc.). Such ambiguous legal statuses inhibited the development of sampling methods based on the total number of transnational migrants in the city. The survey team therefore used a chain method: A well-
respected third party would introduce the interviewer to a respondent and
describe the nature and goals of the survey. Each respondent found through
this initial introduction would in turn indicate several other potential respondents. Among those who were well known and respected in their communities were the leaders and activists of civic organizations for immigrants; individuals who work in the Ukraine UNHCR or participate in the
distribution of UNHCR aid; teachers employed at schools where immigrants' children study; and respected community doctors who work in immigrant districts.

At the same time, the research team attempted to develop a sample with
a representative number of respondents from different countries of origin in
proportion to the number of their compatriots living in the city of Kyiv.
Based on the ethnonational structure of registered foreigners in the city of
Kyiv with refugee status or who were asylum seekers, researchers set goals
for the proportional representation of specific ethnic groups within the pool
of respondents. However, these goals served only as guidelines. Any pre-

planned distribution of respondents based on their countries of origin changed over the course of the survey, depending on the ability of interviewers to establish contact with individual ethnic groups of immigrants. Despite these difficult conditions, the general proportions of the sample remained constant.

The survey of transnational migrants was conducted by means of semistructured personal interviews by the members of the research team. Once respondents agreed, interviews were conducted in their homes. However, most immigrants were not prepared to invite interviewers into their residences. In such cases, interviewers did not insist on conducting the survey in the place of residence. Approximately one in three interviews was conducted in immigrants' residences. Almost all these interviews were with immigrants from Afghanistan, Vietnam, and African countries.

The interviewer filled out the questionnaire during or after the interview based on tape-recorded notes. The primary respondent (the main individual questioned by the interviewer) provided information about absent or underage members of the household and also answered open-ended questions for evaluative purposes. At the same time, the interviewer recorded the responses of other members of the household as well.

In addition to the questionnaire, the interviewer tape-recorded other observations that were typical or, on the contrary, unusual expressions or comments by the respondents. The scope and quality of this additional information depended on many factors: the personality of the respondent, the setting of the interview, and the interviewer's ability to establish a rapport with the respondent. Thus, in some cases there was very little of this kind of information, whereas in others the information allowed the researchers to compile a detailed biography of the respondent and his or her family.

The interviews, which usually lasted an hour and a half to two hours, were conducted both at the respondents' residences and their workplaces (Troieshchynsky, Sviatoshynsky, and Volodymyrsky trading markets), at the UNHCR Reception Center, and the UNHCR's Social Center for Refugee Women and Children. Whenever necessary, an interpreter from the transnational migrant community in question who was fluent in Russian and Ukrainian would be invited to participate in the interview. These were mostly individuals with whom interviews had been conducted earlier.

The team encountered various obstacles during the course of the survey. As might be expected given their undefined legal status (or more accurately, its absence), respondents were not always communicative. The language proficiency of the respondents, and sometimes of the interpreter, determined the level of communication as well. Occasionally, the presence of the interpreter had a negative impact on establishing trust. At the same time, if the interpreter was an individual with whom the respondent had a friendly relationship, his or her participation would spark a detailed, lively conversation.

The gender balance of the sample remained a central concern for the survey team. Only 10 percent of the respondents from transnational migrant communities were women. The Vietnamese community proved to be an exception to this pattern, with female respondents accounting for almost one-third of the survey participants from that community. In most instances, the male head of household proved to be the main interlocutor in family homes, with wives perhaps adding passing comments in some instances. Information obtained through the survey therefore reflects mainly the male view of the immigrant situation. However, the few questionnaires that were filled out during conversations with women reveal views that differed somewhat from those of their male counterparts. For most women, in particular those from Muslim countries, communication was possible only at the UNHCR's Social Center for Refugee Women and Children.

Unfortunately, the survey did not provide exhaustive information about all social groups among transnational migrant communities. A major motivation for meeting with researchers proved to be a desire to describe problems in the hope that the researchers might be able to intervene in some positive manner. Those who agreed to participate were most often individuals without a stable income or clearly defined prospects, and those whose legal status was undefined. Immigrants who are more comfortable in Kyiv— those with legal papers, decent employment, and a defined social status— did not show great interest in the survey. They either declined offers to participate in survey or gave very limited information. Figuratively speaking, respondents comprised mostly the less economically successful members of immigrant communities, and only partially members of their community's "middle class."

## Stage Two

The second stage of the study consisted of a survey of indigenous Kyiv residents. Whereas the first stage of the survey examined issues related to the influx of immigrants to Kyiv and their settlement arrangements, at this stage the research team sought to identify and analyze the attitudes of Kyiv residents toward problems connected with the arrival in Kyiv of people from Asian and African countries.

This portion of the study was conducted by the Kyiv-based sociological company Image Control, which carried out a survey in May 2002 based on a representative survey of Kyiv residents. The sample included 1,000 respondents, and it was formulated according to the criteria of gender, age, and education, taking into account employment status and territorial distribution. Territorial distribution was based on the total population of Kyiv and the population of its districts. According to the Kyiv Municipal Department of Statistics, as of early 2001, 2,606,716 people lived in Kyiv. The source for these data was *Kyiv u tsyfrakh v 2000: Statystychnyi dovidnyk* (Kyiv: Derzhavnyi komitet statystyky Ukrainy, Kyivske miske upravlinnia statystyky, 2001), 159.

The city of Kyiv's population can be broken down by district as follows: Darnytsky, 250,000; Desniansky, 320,000; Dniprovsky, 350,000; Holosiivsky, 200,000; Obolonsky, 290,000; Pechersky, 170,000; Podilsky, 210,000; Shevchenkivsky, 200,000; Solomiansky, 280,000; and Sviatoshynsky, 360,000. Proportional to these data, the number of surveyed individuals by district was Darnytsky, 101; Desniansky, 124; Dniprovsky, 140; Holosiivsky, 78; Obolonsky, 101; Pechersky, 60; Podilsky, 79; Shevchenkivsky, 101, Solomiansky, 100; and Sviatoshynsky, 116.

The survey team selected streets in each Kyiv district for the survey, instructing interviewers not to question more than three respondents from the same building. The route lists included the address and the first and last name of each respondent (if he or she agreed to participate). Refusals to participate in the survey were not recorded.

The survey questions for the second stage of the study focused on how average Kyiv residents view nontraditional immigrants; how they gauge immigrants' opportunities to adapt to life in the city; their attitude toward the newcomers and immigrants' attitudes toward them; and whether Kyiv residents are prepared to live and work alongside people from distant countries. Interviewers separately tape-recorded Kyiv residents' opinions about the newcomers' education, their employment, medical services, and so on. In addition, Kyiv residents evaluated the municipal government's readiness and ability to address immigrants' problems related to their arrival and residence in the city.

## Stage Three

During the third stage of the study, the researchers conducted a survey of individuals whose official duties or type of work involve direct contact with transnational migrants residing in Kyiv, and who therefore have special knowledge and a deeper understanding of the problem than the average Kyiv resident. Among those interviewed were several Ukrainian national government officials who are directly involved in formulating state migration policy; officials of the Kyiv municipal and local administrations; officers of the municipal and local police departments; medical personnel; teachers; and representatives of nongovernmental organizations who work with nontraditional immigrants.

The research teams interviewed forty-six experts using a conversation format based on a previously prepared questionnaire, which included specific questions pertaining to the respondent's area of expertise as well as a set of general questions asked of all the respondents. The questions focused on an analysis of various aspects of the life of nontraditional immigrants in the city of Kyiv and the attitudes of the municipal government and average citizens toward this problem.

The results of this collaborative research effort have been published in several English-language works, including Nancy E. Popson and Blair A. Ruble, "Kyiv's Nontraditional Immigrants," *Post-Soviet Geography and Economics* 41, no. 5 (2001): 365–78; Popson and Ruble, "A Test of Urban Social Sustainability: Societal Responses to Kyiv's 'Non-Traditional' Migrants," *Urban Anthropology* 30, no. 4 (2001): 381–409; Ruble, "Kyiv's Troeshchyna: An Emerging International Migrant Neighborhood," *Nationalities Papers* 31, no. 2 (2003): 139–55; and Olena Braichevska, Halina Volosiuk, Olena Malynovska, Yaroslav Pylynskyj, Nancy Popson, and Blair Ruble, *Nontraditional Immigrants in Kyiv* (Washington, D.C.: Woodrow Wilson Center Comparative Urban Studies Project and Kennan Institute, 2004). More complete findings are to be found in a Ukrainian-language volume: O. Braichevska, G. Volosiuk, O. Malinovska, Ia. Pilinskii, N. Popson, and B. Rubl', *Netraditsiini immigranti u Kyivi* [Nontraditional immigrants in Kyiv] (Kyiv: Kennan Kyiv Project, 2003).

# Notes

## Chapter 1—Creating Diversity Capital: Migrants in Divided Cities

1. Peter Hamill, *Downtown: My Manhattan* (New York: Little, Brown, 2004), 269–70.

2. Thomas Bender, *The Unfinished City: New York and the Metropolitan Idea* (New York: New Press, 2002), 192.

3. Bender, *Unfinished City.*

4. Russell Shorto, *The Island at the Center of the World: The Epic Story of Dutch Manhattan and the Forgotten Colony That Shaped America* (New York: Doubleday, 2004), 2.

5. Bender, *Unfinished City,* 187.

6. Shorto, *Island at the Center of the World,* 265.

7. Shorto, *Island at the Center of the World,* 310.

8. Shorto, *Island at the Center of the World,* 284–311.

9. The concept of "pragmatic pluralism" in an urban setting is developed further in Blair A. Ruble, *Second Metropolis: Pragmatic Pluralism in Gilded Age Chicago, Silver Age Moscow, and Meiji Osaka* (Cambridge: Cambridge University Press, 2001; Baltimore: Johns Hopkins University Press, 2004).

10. For further discussion of the emergence of a culture of forced tolerance in Amsterdam, see Geert Mak, *Amsterdam,* trans. Philipp Blom (Cambridge, Mass.: Harvard University Press, 2000), 97–133.

11. Richard Stren and Mario Polese, "Understanding the New Sociocultural Dynamics of Cities: Comparative Urban Policy in a Global Context," in *The Social Sustainability of Cities: Diversity and the Management of Change,* ed. Richard Stren and Mario Polese (Toronto: University of Toronto Press, 2000), 3–38; the quotation here is on 3.

12. Pierre Bourdieu, "Cultural Reproduction and Social Reproduction," in *Knowledge, Education, and Cultural Change,* ed. R. Brown (London: Tavistock, 1973), 71–112; Pierre Bourdieu, "Symbolic Power," in *Identity and Structure: Issues in the Sociology of Education,* ed. D. Gleason (Dimiffield, U.K.: Nefferton, 1977), 112–19; and Pierre Bourdieu, "The Forms of Capital," in *Handbook of Theory and Research for the*

*Sociology of Education,* ed. J. G. Richardson (Westport, Conn.: Greenwood Press, 1986), 241–58.

13. Robert D. Putnam, *Making Democracy Work: Civic Institutions in Modern Italy* (Princeton, N.J.: Princeton University Press, 1993); Putnam, *Bowling Alone: The Collapse and Revival of American Community* (New York: Simon & Schuster, 2000).

14. For perhaps the most influential article exploring the distinction between forms of social capital and civil society that undermine trust in the relationships between state and society, as opposed to those which reinforce virtuous cycles of interaction, see Simone Chambers and Jeffrey Kopstein, "Bad Civil Society," *Political Theory* 29 (2001): 837–65. For another particularly informative and insightful discussion of the difference between "bonding" and "binding" as used by Putnam for "good" and "bad" social capital, see Annick Germain, "Capital social et vie associative de quartier en contexte multiethnique: Quelques réflexions á partier de recherches montréalaises," *Journal of International Migration and Integration / Revue de l'Integration et de la Migration Internationale* 5, no. 2 (2004): 191–206.

15. For an excellent discussion of such approaches, see the special issue of the Canadian *Journal of International Migration and Integration / Revue de l'Integration et de la Migration Internationale* in which contributors seek to apply social capital perspectives to explore how various forms of social capital "operate in the context of immigration and ethnocultural diversity; to understand the role of social capital in the social, economic, civic, and political participation of immigrants and ethno-cultural groups; and finally to draw lessons for public policy": Jean-Pierre Voyer, "Foreword," *Journal of International Migration and Integration / Revue de l'Integration et de la Migration Internationale* 5, no. 2 (2004): 159–64. The contributors to this special journal issue—which grew out of a conference convened in Montreal in November 2003—include Voyer, Jean Lock Kunz, Peter S. Li, Annick Germain, Annika Forsander, Yvonne Hébert, Xiaochong Shirley Sun, and Eugene Kowch.

16. The flow of people across borders in the present era is exceeded only by the forced intercontinental migration of black slaves from Africa to the Americas and the Caribbean in the eighteenth and nineteenth centuries, and by the exodus of 55 million Europeans to the New World and Australia in the century following 1820. The background for this observation is explored in such works as T. J. Hatton and J. G. Williamson, *The Age of Mass Migration. Causes and Economic Impact* (New York: Oxford University Press., 1998), 3–7; Caroline B. Bretell and James F. Hollifield, eds., *Migration Theory: Talking across Disciplines* (New York: Routledge, 2000); and Douglas S. Massey, Joaquin Arango, Graeme Hugo, Ali Kouaouci, Adela Pellegrino, and J. Edward Taylor, *Worlds in Motion: Understanding International Migration at the End of the Millennium* (Oxford: Clarendon Press, 1998).

17. Sarah Collinson, *Europe and International Migration* (London: Royal Institute of International Affairs, 1993).

18. Elazaer Barkan and Marie-Denise Shelton, *Borders, Exiles, Diasporas* (Stanford, Calif.: Stanford University Press, 1998).

19. The importance of cities in the life experiences of transnational migrants at the beginning of the twenty-first century is explored in Stephen Castles and Mark J. Miller, *The Age of Migration: International Population Movements in the Modern World,* 3rd ed. (New York: Guilford Press, 2003), esp. 228–30.

20. Martin Heisler, "Now and Then, Here and There: Migration and the Transformation of Identities, Borders, and Orders," in *Identities, Borders, and Orders,* ed. Math-

ias Akbert, David Jacobson, and Yosef Lapid (Minneapolis: University of Minnesota Press, 2001), 225–47.

21. The term "transnational" has been an object of considerable debate and discussion, having been subjected to justified criticism as being little more than a capricious renaming of social phenomena that have long been in existence. Though some of these critiques have validity, the term remains most appropriate for the discussion here for a number of reasons.

The term will be used in this volume because, as Nadje Al-Ali and Khalid Koser argue, "transnational" retains two advantages over the more traditional vocabulary surrounding "international" in relation to twenty-first century migration. First, "transnational" reflects a contemporary migration experience which remains dynamic and open rather than static. Migrants in cities such as Montreal, Washington, and Kyiv do not necessarily move from "home" to "host" or "receiving" societies. Instead, residence in any given locale is but one moment in a life course which could well involve further destinations.

Second, "transnational discourse" on migration assumes that the geographical movement of people across international borders has become a "normal state" rather than an aberration. Part of the argument to follow is predicated on the assumption that migrant communities from beyond national boundaries will be a standard feature of everyday life in cities such as Montreal, Washington, and Kyiv long into the future. For further discussion of these issues, please see Nadje Al-Ali and Khalid Koser, "Transnationalism, International Migration and Home," in *New Approaches to Migration? Transnational Communities and the Transformation of Home,* ed. Nadje Al-Ali and Khalid Koser (New York: Routledge, 2002), 1–14.

A third characteristic of contemporary migration experience emphasized by the transnational literatures seems particularly appropriate for understanding the situation in early twenty-first-century Montreal, Washington, and Kyiv: the coexistence of different migration communities with differing levels of commitment to any given place. Transnational migrants who think of their presence at any given location will be short-lived are neighbors with other transnational migrants who are looking to settle in for the *longue durée.* E.g., see the discussion in Nicholas Van Hear, *New Diasporas: The Mass Exodus, Dispersal and Regrouping of Migrant Communities* (London: UCL Press, 1998), 241–56.

"Transnational" also seems especially appropriate for this particular study for a fourth reason, which is explored in detail by Aihwa Ong, among others. Previous understandings of "international migration" and "immigration" have tended to emphasize a hierarchy in which migrants move "up" and "benefit from" their new host societies. The migrants at the center of this study arguably move "among" communities in a network rather than "up" a hierarchy. Their new host cities—Montreal, Washington, and Kyiv—are enriched as much or more by their presence as are the new arrivals themselves. Please see Aihwa Ong, *Flexible Citizenship: The Cultural Logistics of Transnationality* (Durham, N.C.: Duke University Press, 1999).

All these dimensions suggest that the migration experience during the period under review here is different in meaningful ways from what traditionally was thought of as "immigration" a generation or two ago. This point is explored in Saskia Sassen, *Guests and Aliens* (New York: New Press, 1999).

22. Hugh MacLennan, *Two Solitudes* (Markham, Ontario: Fitzhenry & Whiteside, 1996), 3.

23. Landry's efforts to reconstruct Canadian history as a series of negotiations between two of the country's founding groups—French and English—provides a useful and concise overview of this long and complex history. Rejean Landry, "The Political Development of Quebec," in *Identities: The Impact of Ethnicity on Canadian Society,* ed. Wsewolod Isajiw (Toronto: Peter Martin, 1977), 71–84. More encompassing historical overviews may be found in Desmond Morton, *A Short History of Canada,* 3rd rev. ed. (Toronto: McClelland & Stewart, 1997), 24–49; Scott W. See, *The History of Canada* (Westport, Conn.: Greenwood Press, 2001), 39–6; Paul-André Linteau, René Durocher, and Jean-Claude Robert, *Histoire du Québec Contemporain* (Québec: Les Éditions du Boréal Express, 1979), 16–22; and Laurier La Pierre, *Québec: A Tale of Love* (Toronto: Penguin Viking Canada, 2001)—as well as in several works catalogued in Paul Aubin, *Bibliographie de l'histoirie du Québec et du Canada, 1966–1975, Tome 1* (Quebec City: Institut Québeçois de Réchérche sur la Culture, 1981).

24. D. F. Levy and L. S. Bourne, "The Social Context and Diversity of Urban Canada," in *The Changing Social Geography of Canadian Cities,* ed. Larry S. Bourne and David F. Levy (Montreal: McGill–Queen's University Press, 1993), 3–30; the citation here is on 1.

25. A central theme in many works, the importance of the linguistic divide may be immediately discerned from standard histories of the city among many sources. E.g., see Paul-André Linteau, *Histoire de Montréal depuis la Confederation* (Quebec City: Les Éditions du Boréal, 1992), as well as Annick Germain and Damaris Rose, *Montreal: The Quest for a Metropolis* (New York: John Wiley & Sons, 2000).

26. Harold Kaplan, *Reform, Planning and City Politics: Montreal, Winnipeg, Toronto* (Toronto: University of Toronto Press, 1982), 315.

27. Kaplan, *Reform, Planning and City Politics.*

28. Kaplan, *Reform, Planning and City Politics,* 322.

29. The watershed of Quebec's "Quiet Revolution" is discussed in the historical reviews found in the notes above. A brief overview of the importance of this period for language politics may be found in Dominique Arel, "Political Stability in Multinational Democracies: Comparing Language Dynamics in Brussels, Montreal, and Barcelona," in *Multinational Democracies,* ed. Alain-G. Gagnon and James Tully (Cambridge: Cambridge University Press, 2001), 65–89.

30. Marcel Fournier, Michael Rosenberg, and Deena White, "Quebec as a Distinct Society," in *Quebec Society: Critical Issues,* ed. Marcel Fournier, Michael Rosenberg, and Deena White (Scarborough, Ontario: Prentice-Hall Canada, 1997), 1–16; the citation here is on 7.

31. Arel, "Political Stability in Multinational Democracies," 81.

32. Arel, "Political Stability in Multinational Democracies," 77.

33. A point emphasized in Kaplan, *Reform, Planning and City Politics,* 312–13.

34. S. H. Olson and A. I. Kobayashi, "The Emerging Ethnocultural Mosaic," in *Changing Social Geography of Canadian Cities,* ed. Bourne and Levy, 138–52.

35. Germain and Rose, *Montreal,* 215, 220.

36. Michel Paille, *Nouvelles tendances demo-linguistiques dans l'Isle de Montréal, 1981–1996* (Quebec City: Counseil de la Langue Française, 1989), 16, 19.

37. See http://www12.statcan.ca/english/census01/products/highlight/Language Composition/Page.cfm?Lang=E&Geo=CMA&Code=0&View=1a&Table=1a&Start Rec=76&Sort=2&B1=Counts&B2=Both.

38. L'INRS–Urbanisation, Culture et Société, *Portrait des populations immigrante*

*et non immigrante de la ville de Montréal et de ses 27 arrondissements* (Montreal: Bureau des Relations Interculturelles de la Ville de Montréal, 2003); available at http://www2.ville.montreal.qc.ca/ diversite/ portrait.htm.

39. Radio Canada, "Le français comme langue d'usage à la baisse à Montréal," http://www.radio-canada.ca/regions/Montreal/nouvelles/200501/12/006-FRANCAISMTL.shtml?ref=rss.

40. Linteau, *Histoire de Montréal,* 161–66, 317–22, 464–74; Jeffrey G. Reitz, *The Survival of Ethnic Groups* (Toronto: McGraw-Hill Ryerson, 1980), 89.

41. For a discussion of early Italian migration to Montreal in particular, see Bruno Ramirez, "The Crossroad Province: Quebec's Place in International Migrations, 1870–1915," in *A Century of European Migrations, 1830–1930,* ed. Rudolph J. Vecoli and Suzanne M. Sinke (Urbana: University of Illinois Press, 1991), 243–60; the citation here is on 250–54.

42. Mordecai Richler, "Quebec Oui, Ottawa Non," in *Home Sweet Home: My Canadian Album,* ed. Mordecai Richler (New York: Alfred A. Knopf, 1984), 27–45; the quotation here is on 38.

43. MacLennan, *Two Solitudes,* 25.

44. Richler, "Quebec Oui, Ottawa Non," 37.

45. T. R. Balakrishnan, "Changing Patterns of Ethnic Residential Segregation in Metropolitan Areas of Canada," *Canadian Review of Sociology and Anthropology* 19 (1982): 92–110; T. R. Balakrishnan and K. Selvanathan, "Ethnic Segregation in Metropolitan Canada," in *Ethnic Demography,* ed. S. S. Halli, F. Trovato, and L. Driedger (Ottawa: Carleton University Press, 1990), 399–413; W. K. D. Davies and R. A. Murdie, Measuring the Social Ecology of Cities," in *Changing Social Geography of Canadian Cities,* ed. Bourne and Levy, 52–75 (the citation here is on 70–72); and Raymond Breton, Wsewolod W. Isajiw, Warren E. Kalbach, and Jeffrey G. Reitz, *Ethnic Identity and Equality: Varieties of Experience in a Canadian City* (Toronto: University of Toronto Press, 1990).

46. T. R. Balakrishnan and J. Kralt, "Segregation of Visible Minorities in Montreal, Toronto, and Vancouver," in *Ethnic Canada,* ed. L. Driedger (Toronto: Copp Clark Pitman, 1987), 138–57; the citation here is on 145.

47. Germain and Rose, *Quest for a Metropolis,* 214.

48. The formulation in the heading on this section—"From 'Chocolate City, Vanilla Suburbs' to 'Rocky Road'"—was the title of a newspaper column by the author published in the *Washington Afro-American* newspaper. Blair A. Ruble, "From 'Chocolate City, Vanilla Suburbs' to 'Rocky Road,'" *Washington Afro-American,* October 16–22, 2004.

49. The most detailed representation of the area prior to the establishment of Washington may be found in the Fry and Jefferson Map published in the 1750s in London. Various editions of this map over the next quarter-century—prepared by Joshua Fry, a professor of mathematics at William and Mary College, and Peter Jefferson, the father of future President Thomas Jefferson—reveal a number of plantations , towns, and "paper towns" in a region expending from the Atlantic to the East to the Appalachians to the West, from southern Pennsylvania in the North to northern North Carolina in the south. For further discussion of the Fry and Jefferson Map as well as other early maps of the region, see Iris Miller, *Washington in Maps, 1606–2000* (New York: Rizzoli International Publications, 2002), 26–27.

50. A brief overview of the settling of the area that is now Washington may be found in David L. Lewis, *District of Columbia: A Bicentennial History* (Nashville and New

York: American Association for State and Local History and W.W. Norton & Company, 1976), 1–15.

51. Lewis, *District of Columbia*, 5–6. This narrative of the negotiations between the first commissioners and local land owners is taken from Lewis's Washington-oriented account. Other accounts of the bargaining and compromise focus instead on the national issues involved revolving around resolution of debt. For one excellent recent analysis of these negotiations from a national perspective, see Ron Chernow's retelling of the story in *Alexander Hamilton* (New York: Penguin Press, 2004), 310–31.

52. Howard Gillette Jr., *Between Justice and Beauty: Race, Planning, and the Failure of Urban Policy in Washington, D.C.* (Baltimore: Johns Hopkins University Press, 1995), 22–23.

53. Lewis, *District of Columbia*, 7–15.

54. On Banneker's indispensable role in laying out the city, see Lewis, *District of Columbia*, 40–42; as well as Constance McLaughlin Green, *The Secret City: A History of Race Relations in the Nation's Capital* (Princeton, N.J.: Princeton University Press, 1967); Daniel Murray, *Banneker, the Afro-American Astronomer* (Washington, D.C., 1921); and Silvio Bedini, *The Life of Benjamin Banneker* (New York: Charles Scribner's Sons, 1972).

55. Ironically, Banneker is memorialized by a run-down isolated traffic turnaround at the rear of I. M. Pei's cement brutalist L'Enfant Plaza complex, constructed during the late 1960s and early 1970s on the site of Washington's largest slave market.

56. E.g., see the maps found in Miller, *Washington in Maps*. Significantly, in this regard, one of the opening exhibits at the ill-fated City Museum of Washington in 2003 featured exquisite maps of the city from the Albert Small Collection, http://www.city-museumdc.org. The City Museum remained open about a year before falling victim to low attendance and the absence of continued congressional support. Jacqueline Trescott, "City Museum Moves Up Closing Date to Nov. 28," *Washington Post*, November 10, 2004.

57. John Michael Vlach, "Looking behind the Marble Mask: Varied African American Responses to Difficult History in Washington D.C.," in *Composing Urban History and the Constitution of Civic Identities*, ed. John Czaplicka and Blair A. Ruble (Washington and Baltimore: Woodrow Wilson Center Press and Johns Hopkins University Press, 2003), 31–57; the citation here is on 47–48.

58. Gillette, *Between Justice and Beauty*, 30–35.

59. Vlach, "Looking behind the Marble Mask," 52. For further discussion of African-American life in Washington during the Civil War, see Margaret Leech, *Reveille in Washington, 1860–1865* (New York: Carroll & Graf, 1991), 234–58.

60. The history of "Jim Crow" laws in Washington, and their relation to legal racial segregation elsewhere in the United States, are discussed at length in Green, *Secret City*; and Richard L. Kluger, *Simple Justice: The History of Brown v. Board of Education and Black America's Struggle for Equality* (New York: Random House, 1976).

61. Edward P. Jones, "The Store," in *Lost in the City*, ed. Edward P. Jones (New York: HarperPerennial, 1992), 83–110.

62. Jones, "Store," 109–10.

63. James A. Miller, "Black Washington and the New Negro Renaissance," in *Composing Urban History and the Constitution of Civic Identities*, ed. Czaplicka and Ruble, 217–39.

64. Gillette, *Between Justice and Beauty*, 28; Alan Lessoff, *The Nation and Its City:*

*Politics, "Corruption," and Progress in Washington, D.C., 1861–1902* (Baltimore: Johns Hopkins University Press, 1994), 18.

65. Gillette, *Between Justice and Beauty,* 28.

66. Gillette, *Between Justice and Beauty,* 153.

67. For a discussion of Washington alleys as places to live, see James Borchert, *Alley Life in Washington: Family, Community, Religion, and Folklife in the City, 1850–1970* (Urbana: University of Illinois Press, 1980). Also see Blair A. Ruble, "St. Petersburg's Courtyards and Washington's Alleys: Officialdom's Neglected Neighbors," in *Debat de Barcelona (III): Ciutat real, ciutat ideal,* ed. Pep Subiros (Barcelona: Centre de Fultura Contemporània de Barcelona, 1998), 11–27; this paper is also available as Blair A. Ruble, *St. Petersburg's Courtyards and Washington's Alleys: Officialdom's Neglected Neighbors,* Occasional Paper 285 (Washington, D.C.: Kennan Institute, Woodrow Wilson Center, 2003), http://www.wilsoncenter.org/topics/pubs/stpetersburg/pdf.

68. Gillette, *Between Justice and Beauty,* 73–79; Laura Bergheim, *The Washington Historical Atlas: Who Did What When and Where in the Nation's Capital* (Rockville, Md.: Woodbine House, 1992), 239–47; the citation here is on 256–67.

69. Francine Curro Cary, ed., *Urban Odyssey: A Multicultural History of Washington, D.C.* (Washington, D.C.: Smithsonian Institution Press, 1996).

70. Robert Manning, "Multicultural Washington, D.C.: The Changing Social and Economic Landscape of a Post-Industrial Metropolis," *Ethnic and Racial Studies* 21, no. 2 (1998): 328–54.

71. This sad tale is retold in Esther Ngan-ling Chow, "From Pennsylvania Avenue to H Street, NW. The Transformation of Washington's Chinatown," in *Urban Odyssey,* ed. Cary, 190–207.

72. Gillette, *Between Justice and Beauty,* 160. As in South Africa, racial divisions were enforced by the power of law. E.g., the NAACP's lead attorney (and future Supreme Court justice) Thurgood Marshall was forced to eat lunch at a Union Station canteen for railway porters while arguing the landmark 1954 *Brown v. Board of Education* case before the U.S. Supreme Court because he was prohibited by D.C. statute from dining elsewhere on Capitol Hill.

73. Edward Christopher Williams, *When Washington Was in Vogue: A Love Story,* with commentary by Adam McKible and Emily Bernard (New York: Amistad/HarperCollins, 2003), 160. This masterful novel—an African-American *Great Gatsby* of sorts—had been previously published anonymously as *The Letters of Davey Carr* in the *Messenger* between January 1925 and 1926.

74. Carl Abbott, "Dimensions of Regional Change in Washington, D.C.," *American Historical Review* 95, no. 5 (1990): 1367–93.

75. Gillette, *Between Justice and Beauty,* 153; U.S. Bureau of the Census, *Profiles of General Demographic Characteristics: 2000 Census of Population and Housing, District of Columbia* (Washington, D.C.: U.S. Bureau of the Census, 2000).

76. Gillette, *Between Justice and Beauty,* 153; U.S. Bureau of the Census Web site, http://www.census.gov/Press-Release/metro01.prn.

77. A useful discussion of Ukraine's complex religious landscape may be found in Andrew Wilson, *The Ukrainians: Unexpected Nation* (New Haven, Conn.: Yale University Press, 2000), 234–52.

78. Chauncy D. Harris, "The New Russian Minorities: A Statistical Overview," *Post-Soviet Geography* 34, no. 1 (1993): 1–27 (the citation here is on 17–21); and Irina

Pribytkova, "Kiev i Kievliane na poroge dvadtsat' pervogo veka," unpublished paper, January 2000, tablitsa 5, pp. 44–46.

79. Harris, "New Russian Minorities," 18, and Pribytkova, "Kiev i Kievliane."

80. This point is emphasized in the works of Taras Kuzio and Andrew Wilson. E.g., see Nancy Popson, "Meeting Report on Presentation of Taras Kuzio, Nation-Building in Ukraine: A Growing Elite Consensus," *Kennan Institute Meeting Report* 17, no. 5 (2000); and Popson, "Meeting Report on Presentation of Andrew Wilson: Ukrainian National Identity, the 'Other' Ukraine," *Kennan Institute Meeting Report* 17, no. 9 (2000). The issue of Ukrainian identity is hardly unique to Kyiv. The December 2001 national census reported that 77.8 percent of Ukraine's population of 37,541,700 identified themselves as "Ukrainian" (an increase of 5 percent since 1989) while 67.5 percent of the entire population consider Ukrainian to be their "mother tongue" (an increase of 2.8 percent since 1989). Meanwhile, 17.3 percent of Ukraine's population identified themselves as "Russian" (a decrease of 5 percent during the intercensus period), whereas 29.6 percent of people in Ukraine consider Russian to be their "mother tongue" (a decrease of 3.2 percent since 1989. Canadian Institute of Ukrainian Studies, Toronto Office, e-mail communication, December 30, 2002.

81. Oleh Wolownya, "2001 Census Results Reveal Information on Nationalities and Language in Ukraine," *Ukrainian Weekly,* January 12, 2003.

82. Paul Robert Magocsi, *A History of Ukraine* (Seattle: University of Washington Press, 1996), 49–61.

83. Magocsi, *History of Ukraine,* 66–82.

84. Wilson, *Ukrainians,* 42.

85. Magocsi, *History of Ukraine,* 104–13.

86. Michael F. Hamm, *Kiev: A Portrait, 1800–1917* (Princeton, N.J.: Princeton University Press, 1993), xi–xiii.

87. Magocsi, *History of Ukraine,* 206–16; Wilson, *Ukrainians,* 60–71.

88. A history of this period may be found in Timothy Snyder, *The Reconstruction of Nations: Poland, Ukraine, Lithuania, Belarus, 1569–1999* (New Haven, Conn.: Yale University Press, 2003).

89. For more on this debate, see Snyder, *Reconstruction of Nations;* Magosci, *History of Ukraine,* 51–64; and Wilson, *Ukrainians,* 72–100.

90. As may be seen in Timothy Snyder's analysis of these concerns in Snyder, *Reconstruction of Nations.*

91. Hamm, *Kiev,* 10. All three districts are part of today's central Kyiv.

92. Hamm, *Kiev,* 16.

93. Hamm, *Kiev,* 16–17.

94. Hamm, *Kiev,* 10, 104.

95. Hamm, *Kiev,* 104.

96. Hamm, *Kiev,* 173–207.

97. Magosci, *History of Ukraine,* 51–64; and Wilson, *Ukrainians,* 488–520. The chaos of this period in the city's history is captured in Mikhail Bulgakov's novella *The White Guard,* trans. Michael Glenny (Chicago: Academy Publishers, 1984.

98. E.g., see Robert Conquest, *The Harvest of Sorrow: Soviet Collectivization and the Great Famine of 1932–1933* (London: Hutchinson,1986).

99. Magosci, *History of Ukraine,* 51–64; and Wilson, *Ukrainians,* 557–71.

100. Wilson, *Ukrainians,* 631–32.

101. Wilson, *Ukrainians,* 634–37.

102. Wilson, *Ukrainians,* 648–51. This story is also told in Snyder, *Reconstruction of Nations,* 179–214; and, William Taubman, *Khrushchev: The Man and His Era* (New York: W. W. Norton, 2003).

103. Magosci, *History of Ukraine,* 648–51. This story is also told in Snyder, *Reconstruction of Nations,* 179–214; and in Taubman, *Khrushchev.*

104. David R. Marples, "Chernobyl': A Reassessment," *Eurasian Geography and Economics* 45, no. 8 (2004): 588–607.

105. C. J. Chivers, "Premier Claims He's the Winner in Ukraine Vote, but Observers See Fraud. Crowds Protest State Tally—Outcome to Affect Ties With the West," *New York Times,* November 23, 2004.

106. Kennan Institute Survey of Migrant Families, Kyiv, June–November 2001, question 149. The research team members were Olena Braychevska, Olena Malynovska, Nancy Popson, Yaroslav Pylynsky, Blair Ruble, and Galina Volosiuk. The project's results appear in O. Braichevs'ka, G. Volosiuk,.O. Malinovs'ka, Ia. Pilins'kii, N. Popson, and B. Rubl, *"Netraditsiyni" immigranty u Kyivy* (Kyiv: Institut Kennana / Kyiv'skii Proekt, 2003); as well as in Olena Braichevska, Halyna Volosiuk, Olena Malynovska, Yaroslav Pylynsky, Nancy Popson, and Blair A Ruble, *Nontraditional Immigrants in Kyiv* (Washington, D.C.: Woodrow Wilson Center Comparative Urban Studies Project and Kennan Institute, 2004). For more information about this survey, see the appendix to the present volume.

107. The national security implications of Ukraine's unusual status as a country of large-scale out-migration and simultaneous in-migration is explored in Leonid Polyakov, "Illegal Migration: Ukraine," *European Security* 13 (2004): 17–33.

108. Genadiy Genadiyovich Moskal', committee head, State Committee of Ukraine for Nationalities and Migration, interview, Kyiv, September 21, 2004.

109. Serhiy Brytchenko, "Ukrainian Citizenship: Trends and Prospects," *Uryadovyy Kuryer,* August 5, 2004; published as "Migration in the News," BBC Monitoring / BBC Source News Service, August 12, 2004.

110. A. Babiichuk, "Netradytsiini etnichni hromady u velykykh mistakh Ukrainy: realii ta mozhlyvi upravlinski rishennia," *Upravlinnia Suchasnym Mistom* 5, nos. 1–3 (2002): 148–49; and Kyiv u tsyfrakh 2000, *Statysychnyi zbirnyk* (Kyiv: State Committee for Statistics of Ukraine, 2001), 159. For further discussion of the state of data collection concerning transnational migrants in Kyiv, see Braichevska et al., *Nontraditional Immigrants,* 8–10.

111. Anne de Tinguy explores the impact of newly opened borders on the Russian Federation in her study of the movement of people in and out of Russia following the collapse of the Soviet Union: Anne de Tinguy, *La Grande Migration: La Russie et les Russes depuis l'ouverture du rideau de fer* (Paris: Plon, 2004). An overview statement concerning the significance of border openings in post-Soviet states for broader discussions of migration issues is to be found on pages 14–17.

112. Kennan Institute Survey of Migrant Families, questions 3, 45–65, 145–58, 161, 224–25, 261.

113. Kennan Institute Survey of Migrant Families, question 180.

114. The first four metropolitan regions were New York, Los Angeles, Chicago, and Miami. A quarter-of-a-million immigrants from 193 countries and territories moved to the Washington metropolitan area during the 1990s. For further discussion of these

trends, see Audrey Singer, Samantha Friedman, Ivan Cheung, and Marie Price, *The World in a Zip Code: Greater Washington, D.C. as a New Region of Immigration* (Washington, D.C.: Brookings Institution Press, 2001).

115. D'Vera Cohn, "Educated Minorities Flocked to Region: Population Dipped, Report Says," *Washington Post,* October 5, 2003.

116. The Brookings Institution demographer William H. Frey used these trends to designate Washington as a "melting-pot metro." Cohn, "Educated Minorities Flocked to Region."

117. Phoung Ly, "Churches Adopt African Aura," *Washington Post,* December 7, 2003.

118. Metropolitan Washington is becoming a typical metropolitan region along the lines of what Deyan Sudjic has called "The 100 Mile City," an integrated, multicentered semiurban system of linked—and, at times, antagonistic—urban communities that stretches for hundreds of square miles in every direction; Deyan Sudjic, *The 100 Mile City* (San Diego: Harcourt Brace, 1992). Sudjic's work contributes to a larger body of literature on the emergence of new metropolitan regions at the end of the twentieth century. E.g., see Saskia Sassen, *The Global City: New York, London, and Tokyo* (Princeton, N.J.: Princeton University Press, 1991); Mattei Dogan and John C. Kasarda, eds., *The Metropolis Era* (Beverly Hills, Calif.: Sage, 1988); Neal R. Peirce, *Citistates: How Urban America Can Prosper in a Competitive World* (Washington, D.C.: Seven Locks Press, 1993); and Joel Garreau, *Edge City: Life on the New Frontier* (New York: Doubleday, 1991). Much of this work is an outgrowth of a perspective eloquently set forth by Jane Jacobs in her seminal work *Cities and the Wealth of Nation* (New York: Random House, 1984), which argues that metropolitan regions rather than either cities narrowly defined or nations are the appropriate unit for socioeconomic analysis.

119. Singer et al., *World in a Zip Code.*

120. Two histories of Montreal that deal with the immigrant experience are Jean-Claude Marsan, *Montreal in Evolution* (Montreal: McGill–Queen's University Press, 1981); and Germain and Rose, *Montreal.*

121. L. S. Bourne and J. W. Simmons, "The Dynamics of the Canadian Urban System," in *International Handbook of Urban Systems: Studies of Urbanization and Migration in Advanced and Developing Countries,* ed. H. S. Geyer (Northampton, Mass.: Edward Elgar, 2002), 391–418; the citation here is on 404–7. This has been the case with Toronto and Vancouver as well. Most other Canadian cities receive few immigrants.

122. Elke Laur, "Perceptions linguistiques à Montréal, Tome 1 de 2," Ph.D. dissertation in anthropology, University of Montreal, Montreal, 2001, 14–20.

123. This is apparent in such works as Denise Helly, *L'immigration pour quoi faire?* (Montreal: Institut Québécois de Recherche sur la Culture, 1992); and Anne-Marie Seguin and Annick Germain, "The Social Sustainability of Montreal: A Local or a State Matter?" in *Social Sustainability of Cities,* ed. Polese and Stren, 39–67.

124. A casual observation tested by considerable by social science research. E.g., see Balakrishnan, "Changing Patterns of Ethnic Residential Segregation"; Balakrishnan and Kralt, "Segregation of Visible Minorities"; Davies and Murdie, "Measuring the Social Ecology of Cities"; Olson and Kobayashi, "Emerging Ethnocultural Mosaic"; François Vaillancourt, *Langue et disparities de statut economique au Quebec, 1970–1980* (Quebec City: Consiel de La Langue Française, 1988); and Vaillancourt, *Langue et statut economique au Quebec, 1980–1958* (Quebec City: Consiel de la La Langue Française, 1991).

125. P. Wolff, "Haitians and Anglophone West Indians in the Ethnic and Socio-Economic Structure of Montreal, *Marburger Geographische Schriften* 96 (1984): 286–301.

126. Christopher McAll, "The Breaking Point: Language in Quebec Society," in *Quebec Society,* ed. Fournier, Rosenberg, and White, 61–80.

## Chapter 2—Living in the Middle

1. The increasing importance throughout the 1990s of suburbs as a primary venue for transnational migrants in the life of North American cities is explored by Caroline Brettell, among many authors. E.g., see Caroline Brettell, "Immigrants in an Emerging Gateway City: A Case Study and Some Research Questions," unpublished paper presented at the Woodrow Wilson Center Comparative Urban Studies Project Seminar "Space and Identity: Concepts of Immigration and Integration in Urban Areas," Washington, January 27–28, 2005.

2. Peter S. Li, "Deconstructing Canada's Discourse of Immigrant Integration," *Journal of International Migration and Integration / Revue de l'Intégration et de la Migration Internationale* 4, no. 3 (2003): 315–33.

3. P. Wolff, "Haitians and Anglophone West Indians in the Ethnic and Socio-Economic Structure of Montreal," *Marburger Geographishche Schriften* 96 (1985): 286–301; the quotation is on 286.

4. The 2001 census figures are available on line on the Statistics Canada Web site at http://www.statscan.ca. These figures represent an increase from 1986, when 36.0 percent of the metropolitan Toronto, 28.4 percent of the metropolitan Vancouver, 18.4 percent of metropolitan Montreal, and 2.3 percent of the Quebec City regional population were recorded as being foreign born. For the 1986 figures, see D. F. Levy and L. S. Bourne, "The Social Context and Diversity of Urban Canada," in *The Changing Social Geography of Canadian Cities,* ed. Larry S. Bourne and David F. Levy (Montreal: McGill–Queen's University Press, 1993), 6–7.

5. Ayéko A. Tossou, "Apport Démographique de l'Immigration Internationale dans la Région Métropolitaine de Recensement (RMR) de Montréal, 1976-1996," *Journal of International Migration and Integration / Revue de l'Intégration et de la Migration Internationale* 4, no. 1 (2003): 51–77.

6. Montreal has remained a major magnet for transnational migrants. Overall, 73 percent of the 1.8 million transnational migrants who arrived in Canada between 1991 and 2001 settled in one of three cities: Toronto, Montreal, and Vancouver. Marina Jiménez and Kim Lunman, "Canada's Biggest Cities See Influx of New Immigrants," *Globe and Mail* (Toronto), August 19, 2004.

7. Demaris Rose and Brian Ray, "The Housing Situation of Refugees in Montreal Three Years after Arrival: The Case of Asylum Seekers Who Obtained Permanent Residence," *Journal of International Migration and Integration / Revue de l'Integration et de la Migration Internationale* 2, no. 4 (2001): 493–529; the data are on 494.

8. "Migration Magnet," *Montreal Gazette,* August 5, 2003; Canadian Press, "Montreal Passes Vancouver as Immigrant Destination," *Globe and Mail* (Toronto), August 3, 2003.

9. "Migration Magnet"; Canadian Press, "Montreal Passes Vancouver as Immigrant Destination."

10. This was a point made in a number of studies released at the time, as reported in Radio-Canada, "L'immigration: apport essentiel à l'économie montréalaise," available at http://radio-canada.ca/regions/Montreal/nouvelles/20312/20/004-immigration-travail.sht/.

11. From Statistics Canada Web site, http://www.statcan.ca.

12. Michel Paillé, *Nouvelles tendances démolinguistiques dans l'Isle de Montrèal, 1981–1996* (Quebec City: Conseil de la Langue Française, 1989), 118–19.

13. Bruno Ramirez, "The Crossroad Province: Quebec's Place in International Migrations, 1870–1915," in *A Century of European Migrations, 1830–1930*, ed. Rudolph J. Vecoli and Suzanne M. Sinke (Urbana: University of Illinois Press, 1991), 243–60.

14. Paillé, *Nouvelles tendances démolinguistiques*, 1–2, 20–21, 32, 35.

15. Réne Lévesque, *An Option for Quebec* (Toronto: McClelland and Steward, 1968), 93–94.

16. Claude Bélanger, "Birth Rate," Quebec History Web site, Marianopolis College, http://www2.marianopolis.edu/quebechistory/events/birth8.htm. This decline would only accelerate in the 1990s, with the birthrate falling throughout Quebec for every year after 1990 except for 2000; "Quebec Birthrate Continues to Decline," CBC Montreal Web site, http://montreal.cbc.ca/regional/servlet/View?filename=qc_birth20040420.

17. Levy and Bourne, "Social Context and Diversity," 23. For more extensive discussion of the Asian immigrant experience in Canada, see the essays in *The Silent Debate: Asian Immigration and Racism in Canada*, ed. Eleanor Laquian, Aprodicio Laquian, and Terry McGee (Vancouver: Institute of Asian Research, University of British Columbia, 1998).

18. Denis Helly, *L'immigration pour quoi faire* (Montreal: Institut Québécois de Recherche sur la Culture, 1992).

19. L'INRS–Urbanisation, Culture et Société, *Portrait des populations immigrante et non immigrante de la ville de Montréal et de ses 27 arrondissements*, http://www2.ville.montreal.qc.ca/diversite/portrait.htm. The Canadian Census identified Canadians of "Black, South Asian, Chinese, South East Asian, Arab and Middle Eastern, Latin American, Korean, Japanese, and Filipino heritage" as constituting the country's "visible minorities."

20. This point is explored in Elke Laur, "Perceptions linguistiques à Montréal, Tome 1 de 2," Ph.D. dissertation, Department of Anthropology, Faculty of Arts and Science, University of Montreal, Montreal, May 2001, 41–49, 198–201.

21. Jean-Claude Marsan, *Montreal in Evolution* (Montreal: McGill–Queen's University Press, 1981), 337.

22. Paillé, *Nouvelles tendances démolinguistiques*, 16–23.

23. Anne-Marie Seguin and Annick Germain, "The Social Sustainability of Montreal: A Local or a State Matter?" in *The Social Sustainability of Cities*, ed. Richard Stren and Mario Polese (Toronto: University of Toronto Press, 2000), 39–67; the quotation here is on 42.

24. David Ley, "The New Middle Class in Canadian Central Cities," in *City Lives and City Forms: Critical Research and Canadian Urbanism*, ed. Jon Caulfield and Linda Peake (Toronto: University of Toronto Press, 1996), 19–22.

25. Seguin and Germain, "Social Sustainability of Montreal," 43.

26. Kazi Stastna, "Exploring the Links between Family Makeup, Economy," *Montreal Gazette*, February 11, 2004.

27. Kazi Stastna, "We're Nicely Educated but Poor," *Montreal Gazette,* February 10, 2004.

28. As reported on Radio-Canada/Montreal, "La pauvreté frappe durement les jeunes enfants," available at http://radio-canada.ca/regions/ montreal/version_imprimable asp?nv=/regions/Montreal/, 12/19/2004.

29. Ley, "New Middle Class in Canadian Central Cities."

30. Rose and Ray, "Housing Situation of Refugees in Montreal Three Years after Arrival."

31. Navjot K. Lamba and Harvey Krahn, "Social Capital and Refugee Resettlement: The Social Networks of Refugees in Canada," *Journal of International Migration and Integration / Revue de l'Intégration et de la Migration Internationale* 4, no. 3 (2003): 313–60.

32. Côte-des-Neiges is the subject of extensive social science research in part because of its diverse and complex character and, in part, because of the presence of the University of Montreal and its faculty and students. An excellent collection of articles examining various aspects of neighborhood life during the 1990s appeared in 1997: Deidre Meintel, Victor Piché, Danielle Juteau, and Sylvie Fortin, eds., *Le Quartier Côtes-des-Neiges à Montréal. Les interfaces de la pluriethnicité* (Montréal: L'Harmattan, 1997).

33. Myriame El Yamani, with the assistance of Jocelyne Dupuis, " La construction mediatique du 'Bronx' de Montréal," in *Quartier Côtes-des-Neiges à Montréal,* ed. Meintel et al., 29–52.

34. Bruno Ramirez, "Histoire et histoires dans la métropole québécoise: le quartier Côtes-des-Neiges," in *Quartier Côtes-des-Neiges à Montréal,* ed. Meintel et al., 53–76.

35. Ramirez, "Histoire et histoires dans la métropole québécoise."

36. Jean Renaud and Pierre Legendre, "Nouveaux immigrants et localisation résidentielle," in *Quartier Côtes-des-Neiges à Montréal,* ed. Meintel et al., 103–27; Pierre Joseph Ulysse and Christopher McAll, "Minorités, majorités et les territoires du quotidien," in *Quartier Côtes-des-Neiges à Montréal,* ed. Meintel et al., 191–209; Hélène Bertheleu and Pierre Billion, "Cloisonnement ethnique et solidarité caotive: Familles lao dan le quartier Côtes-des-Neiges," in *Quartier Côtes-des-Neiges à Montréal,* ed. Meintel et al., 299–361; and Sylvie Fortin, "Les Libanais dèimmigration récente: insertion ou exlusioné," in *Quartier Côtes-des-Neiges à Montréal,* ed. Meintel et al., 263–88.

37. Victor Piché and Pierre Legendre, "Le quartier Côtes-des-Neiges: Fiction statistique ou milieu d'insertion pour les groupes d'immigrants," in *Quartier Côtes-des-Neiges à Montréal,* ed. Meintel et al., 77–101; L'INRS–Urbanisation, Culture et Société, *Portrait des populations immigrante et non immigrante de la ville de Montréal.*

38. L'INRS–Urbanisation, Culture et Société, *Portrait des populations immigrante et non immigrante de la ville de Montréal.*

39. L'INRS–Urbanisation, Culture et Société, *Portrait des populations immigrante et non immigrante de la ville de Montréal.*

40. L'INRS–Urbanisation, Culture et Société, *Portrait des populations immigrante et non immigrante de la ville de Montréal.*

41. Daniel Juteau and Sylvie Paré, "L'entrepreneurship à Côtes-des-Neiges: Le périmètre Victoria / Van Horne," in *Quartier Côtes-des-Neiges à Montréal,* ed. Meintel et al., 129–60.

42. Yamani, " La construction mediatique du 'Bronx,'" 35. For a broader discussion of Montreal's "ethnic" press, see Sylvie St-Jacques, "Des nouvelles de leurs mondes," *La Presse,* April 21, 2004.

43. Yamani, "La construction mediatique du 'Bronx'"; and Gladys Symons with Marie-Josée Loiselle, "Le contrôle social et la construction de l'Autre: La police dans un quartier multiethnique," in *Quartier Côtes-des-Neiges à Montréal*, ed. Meintel et al., 173–89.

44. Josiana Le Gall and Deidre Meintel, "Escapes observés: Ethnicité et appropriation territoriale," in *Quartier Côtes-des-Neiges à Montréal*, ed. Meintel et al., 211–28; Ida Simon-Barouh, "Rapports sociaux-ethniques à Côtes-des-Neiges, un quartier de Montréal," in *Quartier Côtes-des-Neiges à Montréal*, ed. Meintel et al., 289–313; and Deirdre Meintel, with Victor Piché and Sylvie Fortin, "Étudier la pluriethnicité à l'ère de la mondialisation," in *Quartier Côtes-des-Neiges à Montréal*, ed. Meintel et al., 11–28.

45. Stephen Henighan, *The Streets of Winter* (Saskatoon: Thistledown Press, 2004), 69.

46. Wolff, "Haitians and Anglophone West Indians."

47. Annick Germain and Damaris Rose, *Montreal: The Quest for a Metropolis* (New York: John Wiley & Sons, 2000), 215.

48. Christopher McAll, "The Breaking Point: Language in Quebec Society," in *Quebec Society: Critical Issues*, ed. Marcel Fournier, Michael Rosenberg, and Deena White (Scarborough, Ont.: Prentice-Hall Canada, 1997), 61–80. A francophone perspective on this debate may be found in Paillé, *Nouvelles tendances démolinguistiques*.

49. These events are the subject of the articles contained in *Trudeau's Shadow. The Life and Legacy of Pierre Elliott Trudeau*, ed. Andrew Cohen and J. L. Granatstein (Toronto: Vintage Canada, 1998).

50. One brief overview of this complex political and legislative history may be found in Helly, *L'immigration pour quoi faire*, 13–24; and in Groupe de Recherche Ethnicité et Société, "Immigration and Ethnic Relations in Quebec: Pluralism in the Making," in *Quebec Society*, ed. Fournier, Rosenberg, and White, 96–105.

51. For more about PQ founder René Lévesque and his time in office, see Graham Fraser, *René Lévesque and the Parti Québécois in Power* (Montreal: McGill–Queen's University Press, 1984).

52. Fraser, *René Lévesque*, 91–112.

53. More precisely, Bill 101 specified that only children with both parents having been schooled in English in Quebec could attend English language schools. This clause was eventually modified by the Canadian Supreme Court, which substituted "in Canada" for "in the Province of Quebec." In the end, the children of transnational migrants had no choice but to attend French language school. Germain and Rose, *Montreal*, 230–31.

54. Deidre Meintel, "Les comportements linguistiques et la nouvelle pluriethnicité Montréalaise," *Études Canadiennese / Canadian Studies* 45, no. 1 (1998): 83–93.

55. Seguin and Germain, "Social Sustainability of Montreal," 48.

56. Dominique Arel, "Political Stability in Multinational Democracies: Comparing Language Dynamics in Brussels, Montreal, and Barcelona," in *Multinational Democracies*, ed. Alain-G. Gagnon and James Tully (Cambridge: Cambridge University Press, 2001), 65–89.

57. Arel, "Political Stability in Multinational Democracies," 75.

58. Arel, "Political Stability in Multinational Democracies," 87.

59. Arel, "Political Stability in Multinational Democracies," 87.

60. Laura-Julie Perreault, "Embouteillage sur le prie-dieu montréalais," *La Presse*, June 12, 2004.

61. Ann Carroll, " Faithful Flock to See Virgin Mary's Tears of Oil," *Montreal Gazette,* February 28, 2004.

62. Seguin and Germain, "Social Sustainability of Montreal," 50.

63. Karen Solie, "The Many Solitudes of Montreal," *Globe and Mail,* June 12, 2004.

64. Hugh MacLennan, *Two Solitudes* (Markham, Ont.: Fitzhenry & Whiteside, 1996).

65. U.S. Bureau of the Census, *Profiles of General Demographic Characteristics: 2000 Census of Population and Housing, District of Columbia* (Washington, D.C.: U.S. Bureau of the Census, 2000).

66. The data mentioned here are from Krishna Roy, "Socio-Demographic Profile," in *The State of Latinos in the District of Columbia,* ed. Council of Latino Agencies / Consejo de Agencias Latinas (Washington, D.C.: Council of Latino Agencies / Consejo de Agencias Latinas, 2002), 1–30. The data are also from Frank Hobbs and Nicole Stoops, U.S. Bureau of the Census, Census 2000, Special Reports, Series CENSR-4, *Demographic Trends in the 20th Century* (Washington, D.C.: U.S. Government Printing Office, 2002), appendix tables 8 and 9. Such data only present general trends, as several observers of the Washington scene warn about the high probability of Census undercounts of the city's black and Latino communities in particular.

67. This account, unless otherwise noted, is based on Howard Gillette Jr., *Between Justice and Beauty: Race, Planning and the Failure of Urban Policy in Washington, D.C.* (Baltimore: Johns Hopkins University Press, 1995), 151–69.

68. As quoted in Gillette, *Between Justice and Beauty,* 163, drawing on a paper by Keith Melder, "In the Capitol's Shadow: Two Neighborhoods," an unpublished manuscript dated August 1978.

69. From a 1957 editorial, as quoted in Gillette, *Between Justice and Beauty,* 163.

70. Zachary M. Schrag, "The Freeway Fight in Washington, D.C.: The Three Sisters Bridge in Three Administrations," *Journal of Urban History* 30, no. 5 (2004): 648–73.

71. David L. Lewis, *District of Columbia: A Bicentennial History* (Nashville and New York: American Association for State and Local History and W. W. Norton, 1976), 185–87.

As will be discussed in chapter 5, District of Columbia "home rule" was limited, representing an uneasy compromise by President Lyndon Johnson and congressional leaders seeking to stabilize what they perceived as growing racial tensions within Washington without turning over final authority to the city's "irresponsible" (i.e., predominantly African-American) residents. Just how limited local authority would prove to be was driven home by a congressionally appointed Financial Control Board, which had oversight of city affairs between 1995 and 2001. Jonetta Rose Barras, *The Last of the Black Emperors: The Hollow Comeback of Marion Barry in the New Age of Black Leaders* (Baltimore: Bancroft Press, 1998), 179–208.

72. Schrag, "Freeway Fight in Washington."

73. Gillette, *Between Justice and Beauty,* 163–64. Also see Harry S. Jaffe and Tom Sherwood, *Dream City: Race, Power and the Decline of Washington, D.C.* (New York: Simon & Schuster, 1994), 29.

74. This boundary was racial rather than economic, because many wealthy African Americans living along the Sixteenth Street, Northwest, corridor in neighborhoods known as the "Gold Coast" had incomes equal to and above those of many whites across the park.

75. Lewis, *District of Columbia,* 196.

76. U.S. Bureau of the Census, http://www.census.gov/population/cencounts/md190090.txt.

77. D.C. Agenda, ed., *2004 Issue Scan: Full Report—An Annual Report Examining Changes in Neighborhood Conditions Across the District of Columbia* (Washington, D.C.: D.C. Agenda, 2004), 112–13.

78. Robert D. Manning, "Washington, D.C.: The Changing Social Landscape of the International Capital City," in *Origins and Destinies. Immigration, Race, and Ethnicity in America,* ed. Silva Pedraza and Rubert G. Rumbaut (New York: Wadsworth Publishing Company, 1996), 384; U.S. Bureau of the Census, http://factfinder.census.gov.

79. Manning, "Washington," 385.

80. D.C. Agenda, *2004 Issue Scan,* 81, 112–13.

81. Between a quarter and a third of residents in Kalorama Heights / Adams Morgan, Columbia Heights / Mount Pleasant are foreign born, with subdistricts within Adams Morgan and Mount Pleasant exceeding these levels. D.C. Agenda, *2004 Issue Scan.*

82. "Executive Summary," in *State of Latinos in the District of Columbia,* ed. Council of Latino Agencies / Consejo de Agencias Latinas, 5.

83. Lois Athey, "Housing," in *State of Latinos in the District of Columbia,* ed. Council of Latino Agencies / Consejo de Agencias Latinas, 73.

84. Athey, "Housing," 74; Debbi Wilgoren, "A Core of City, Old-Timers, Recent Arrivals Seek a New Balance: Population Shifts Leave Some Poorer Residents Worried," *Washington Post,* July 10, 2003.

85. "Executive Summary," in *State of Latinos in the District of Columbia,* 5.

86. Manning, "Washington," 386.

87. Heather McClure, "Child and Family Health," in *2004 Issue Scan,* ed. D.C. Agenda, 18.

88. Manning, "Multicultural Washington," 348.

89. Roy, "Socio-Demograpohic Profile," 3.

90. Mary Beth Sheridan, "Foreign-Born Faring Worse in D.C. Than in Suburbs," *Washington Post,* June 19, 2003.

91. A national pattern identified in detail in Caroline B. Brettell, "Bringing the City Back In: Cities as Contexts for Immigration Incorporation," in *American Arrivals: Anthropology Engages the New Immigration,* ed. Nancy Foner (Santa Fe and Oxford: School of American Research Press and James Currey, 2003), 163–95.

92. Audrey Singer, Samantha Friedman, Ivan Cheung, and Marie Price, *The World in a Zip Code: Greater Washington, D.C. as a New Region of Immigration* (Washington, D.C.: Brookings Institution Press, 2001).

93. Singer et al., *World in a Zip Code,* "Executive Summary."

94. Singer et al., *World in a Zip Code,* 1.

95. Singer et al., *World in a Zip Code,* 3–5.

96. Singer et al., *World in a Zip Code,* 5.

97. Richard Stren and Mario Polese, "Understanding the New Sociocultural Dynamics of Cities: Comparative Urban Policy in a Global Context," in *The Social Sustainability of Cities,* ed. Richard Stren and Mario Polese (Toronto: University of Toronto Press, 2000), 3–38; the quotation here is on 3.

98. Singer et al., *World in a Zip Code,* 11.

99. Singer et al., *World in a Zip Code,* 10–12.

100. Silva Moreno, "At Home with New Opportunities: On 14th Street, Latinos Find Housing, Business Costs They Can Live With," *Washington Post,* March 28, 2002.

101. Darryl Fears, "People of Color Who Never Felt They Were Black: Racial Labels Surprises Many Latino Immigrants," *Washington Post,* December 26, 2002.

102. Darryl Fears, "A Diverse—and Divided—Black Community: As Foreign-Born Population Grows, Nationality Trumps Skin Color," *Washington Post,* February 24, 2002.

103. Fears, "Diverse—and Divided—Black Community."

104. Part of this discussion has appeared previously in Nancy E. Popson and Blair A. Ruble, "Kyiv's Nontraditional Immigrants," *Post-Soviet Geography and Economics* 41, no. 5 (2001): 365–78. For an overview of Soviet-era migration patterns in Ukraine more generally, see A. U. Khomra, "Migratsiia naseleniia Ukrainy v 1959–1988 gg.: Osnovnye zakonnomernosti i tendentsii razvitiia," in *Migratsiia i urbanizatsiia naseleniia (na materialakh Ukrainy i Pol'shii): Sobornik nauchnykh trudov,* ed. A. U. Khomra and P. Eberkhardt (Kyiv: Narodnova Dumka, 1992), 33–52; P. I. Pustokhov, "Dinamika urbanizirovannosti Ukrainy v 1897–1939 gg.," in *Migratsiia i urbanizatsiia naseleniia,* ed. Khomra and Eberkhardt, 74–82; G. L. Glukhanova, "Demograficheskie aspekty urbanizatsii Ukrainskoi SSR v 1959–1989 gg.," in *Migratsiia i urbanizatsiia naseleniia,* ed. Khomra and Eberkhardt, 83–91; and Tat'iana Petrova, *Mikhanizm migratsionnogo obmena: metody issledovaniia* (Kyiv: Naukovo Dumka, 1992).

105. Irina Pribytkova, "Kiev i Kievliane na poroge dvadtsat' pervogo veka," unpublished paper, January 2000, 3–4.

106. Pribytkova, "Kiev i Kievliane."

107. Pribytkova, "Kiev i Kievliane," 5.

108. For a discussion of the repatriation of ethnic Ukrainians, Ukrainian nationals, and their descendants during the 1990s, see Richard H. Rowland, "Urbanization in Ukraine during the 1990s," *Post-Soviet Geography and Economics* 41, no. 3 (2000): 183–216; Olena Malynovska, "Repatriation to Ukraine," *Migration Issues* (Kyiv), no. 4 (1999): 17–29; and Olena Braychevska, "Repatriates in Ukraine: Steps towards Integration," *Migration Issues* (Kyiv), no. 10 (1999): 34–39.

109. Pribytkova, "Kiev i Kievliane," 12–13.

110. Pribytkova, "Kiev i Kievliane," 12–13.

111. Leonid Polyakov, "Illegal Migration: Ukraine," *European Security* 13 (2004): 17–33.

112. Oleksandr Piskun, Irina Prybytkova, and Volodymyr Volovych, *Migratsiina sytuatsii v Ukrayini* (Kyiv: Politicynya Dumka, 1996).

113. This categorization of Kyiv's migrant community—which is discussed further in Popson and Ruble, "Kyiv's Nontraditional Immigrants"—is based on Tomas Frejka, Marek Okolski, and Keith Sword, *In-Depth Studies on Migration in Central and Eastern Europe: The Case of Ukraine* (New York and Geneva: U.N. Economic Commission for Europe and U.N Population Fund, 1999); T. Klinchenko, O. Malynovska, I. Mingazutdinov, and O. Shamshur, *Country Studies on Migrant Trafficking and Alien Smuggling: The Case of Ukraine* (Geneva: International Organization for Migration, 1999); and T. Klinchenko, O. Malynovska, I. Mingazutdinov, and O. Shamshur, *Migrant Trafficking in Ukraine: Report to the International Organization for Migration* (Geneva: International Organization for Migration, 2000).

114. For discussion of interviews with migrants concerning motivation for travel to Ukraine, see Klinchenko et al., *Country Studies on Migrant Trafficking,* 35–38; and

Volodymyr Volovich, Volodymyr Yevtukh, Vladimir Popovich, Vasiliy Kousherec, and Konstantin Korzh, *Transit Migration in Ukraine* (Geneva: International Organization for Migration, 1994), 12–13.

115. For discussion of travel routes to Ukraine, see Klinchenko et al., *Country Studies on Migrant Trafficking;* Klinchenko et al., *Migrant Trafficking in Ukraine;* and Volovich et al., *Transit Migration in Ukraine.* These published analyses of migrant motivations were corroborated by interviews with Ukrainian and Kyiv municipal officials; interviews by the author with Vasyl' Gazhaman, former official, Kyiv City Department of Refugee Status and Migration, Kyiv, January 26, 2000; Mykola Pilipchuk, chief, Administration for Passport, Registration, and Migration Work, Ukrainian Ministry of Internal Affairs, Kyiv, June 30, 2000; Volodymyr Novik, head, Department for Refugees and Minorities, Kyiv City State Administration, Kyiv, June 29, 2000, and May 24, 2001; and Yurii Buznicki, director, Center for Investigations of Migration Problems, Kyiv, June 29, 2000.

116. This is as evidenced in the fact noted in the previous chapter that more new residents in Ukraine requested citizenship in 2003 than native-born Ukrainians officially left the country. Personal communication from Nikolai Shul'ga, deputy director of the Institute of Sociology of the National Academy of Sciences of Ukraine, Kyiv, January 21, 2004.

117. Kennan Institute Survey of Migrant Families, Kyiv, June–November 2001, questions 3, 25, 45–65, 261. For more information about this survey, see the appendix to the present volume.

118. For further discussion of the development of Troeshchyna as Kyiv's leading multiethnic migrant community, see Blair A. Ruble, "Kyiv's Troeshchyna: An Emerging International Migrant Neighborhood," *Nationalities Papers* 31, no. 2 (2003): 139–55.

119. For discussion of late-Soviet-era housing policies, see Blair A. Ruble, "From *Khrushcheby* to *Korobki*," in *Russian Housing in the Modern Age: Design and Social History,* ed. William Craft Brumfield and Blair A. Ruble (Washington and Cambridge: Woodrow Wilson Center Press and Cambridge University Press, 1993), 232–70. Also see A. J. DiMaio, *Soviet Urban Housing: Problems and Policies* (New York: Praeger, 1974); Henry W. Morton, "Who Gets What, When and How? Housing in the Soviet Union," *Soviet Studies* 32 (1980): 235–39; Gregory D. Andrusz, *Housing and Urban Development in the USSR* (Albany: State University of New York Press, 1985); and Carol Nechemias, "The Impact of Soviet Housing Policy on Housing Conditions in Soviet Cities: The Uneven Push from Moscow," *Urban Studies* 18 (1981): 1–8.

120. M. M. Shul'kevich and T. D. Dmitrenko, *Kiev: Arkhiteturnye-istoricheskie ocerhki* (Kyiv: Budivel'nik, 1982), 385–90.

121. Shul'kevich and Dmitrenko, *Kiev,* 390.

122. See, e.g., M. D'iomin, "Stolytsia s'ohodni i zavtra," *Kul'tura i zhyttia,* July 31, 1981, 3; "Aktual'ne interv'iu: Misto soniachnykh kvartaliv," *Vechyrnyi Kyiv,* August 7, 1981, 2; P. Motliakh, "Znyva budivel'ni," *Vechyrnyi Kyiv,* June 16, 1981, 1; V. Koval', "Talant arkhitektora i krasa Kyiva," *Vechyrnyi Kyiv,* May 21, 1981, 1; B. Omel'chenko and I. Sirenko, "Z pohliadom u perspektyvu," *Vechyrnyi Kyiv,* December 17, 1982, 4; P. Motliakh, D. Kanevs'kyi, "Obnovy Mista," *Vechyrnyi Kyiv,* December 31, 1982, 3.

123. I. A. Pravnichenko, "Obraztsovo-pokazatel'nyi mikroraion zhilogo massiva Troeshchina," *Stroitel'stvo i arkhitektura,* no. 11 (1985): 13–15; T. Iu. Vlasova, "Troeshchina, Standart i Rasnoobrazie," *Stroitel'stvo i arkhitektura,* no. 8 (1986):

11–13; I. A. Pravnicheko, "Tsvetnoi mikroraion," *Stroitel'stvo i arkhitektura,* no. 4 (1987): 2–3.

124. Oleksii Plaksin, "Ne skoro potrapysh do spravzhn'oho rynku cherez Kyivs'ki rynky," *Vechyrnyi Kyiv,* October 10, 1995, 3; Iurii Koprnev, "'Merkurii': My nichego ne obeshchaem, my rabotaem vmeste s Vami . . . ," *Kievskie Vedomosti,* February 16, 1995, 4; Vera Frolova, Mikhail Dikalenko, Martovskaia 'sterilizatsiia' rynka," *Kievskie Vedomosti,* March 26, 1996, 25; Mikhail Ostankov, "Bazar, stoi! Raz, dva!" *Biznes,* April 9, 1996, 23; and Genrikh Sikorskii, "Poiavitsia li v Kieve tsivilizovannyi avtorynok?" *Kievskie Vedomosti,* April 20, 1995, 8.

125. "V stolitse rastaskivaiut zemli, a khoziain Kievsoviet—i ne vedaet o tom," *Kievskie Vedomosti,* June 27, 1995, 1, 8.

126. Vitalii Riaboshapka, "S segodniashnego dnia Troeshchinskii rynok prekrash-chaet rabotu: Torgovtsy otkazyvaiutsia pisat' zaiavleniia o poborakh," *Kievskie Vedomosti,* August 13, 1998, 4.

127. Kennan Institute Survey of Migrant Families, questions 45–65, 145–48, 261.

128. Kennan Institute Survey of Migrant Families, questions 45–65, 156–61, 261.

129. Kennan Institute Survey of Migrant Families.

130. "Policies and institutions that have the overall effect of integrating diverse groups and cultural practices in a just and equitable fashion." Stren and Polese, "Understanding the New Sociocultural Dynamics of Cities."

131. For further discussion of this point, see Viktor Stepanenko, "Identities and Language Politics in Ukraine: The Challenges of Nation-State Building," in *Nation-Building, Ethnicity and Language Politics in Transition Countries,* ed. Fariman Daftary and François Grin (Budapest: Open Society Institute Local Government and Public Service Reform Initiative / European Centre for Minority Issues, 2003), 109–35.

132. As based on Richard Stren and Mario Polese's definition of "urban social sustainability"; Stren and Polese, "Understanding the New Sociocultural Dynamics of Cities."

## Chapter 3—Working and Studying in the Middle

1. Ursula Barnes, "Every Colour Under the Sun," in *Diaspora City: The London New Writing Anthology* by Ursula Barnes (London: Arcadia, 2003), 192.

2. Richard Wright and Serin Houston, "Strange Bedfellows? Performity, Mixed Households, and the Negotiation of Urban Space," unpublished paper presented at Woodrow Wilson Center Comparative Urban Studies Project Seminar, "Space and Identity: Concepts of Immigration and Integration in Urban Areas," Washington, January 27–28, 2005. Also, T. Wright, S. Houston, M. Ellis, S. Holloway, and M. Hudson, "Crossing Racial Lines: Geographies of Mixed-Race Partnering and Multiraciality in the United States," *International Journal of Population Geography* 5 (2003): 31–48.

3. Richard Stren and Mario Polese, "Understanding the New Sociocultural Dynamics of Cities: Comparative Urban Policy in a Global Context," in *The Social Sustainability of Cities,* ed. Richard Stren and Mario Polese (Toronto: University of Toronto Press, 2000), 3–38; the quotation here is on 3.

4. Elke Laur, "Perceptions linguistiques à Montréal: Tome 1 de 2," Ph.D. dissertation in anthropology, University of Montreal, 2001, 50–54.

5. Laur, "Perceptions linguistiques à Montréal," 2.

6. This discussion is based on Christopher McAll's important 1997 essay, "The Breaking Point: Language in Quebec Society," which appeared in *Quebec Society. Critical Issues,* ed. Marcel Fournier, Michael Rosenberg, and Deena White (Scarborough, Ont.: Prentice-Hall Canada, 1997), 61–80.

7. McAll, "Breaking Point," 68.

8. Immigrant men constituted 11.0 percent of Montreal's employed labor force in 1996, while immigrant women were 8.3 percent of employed Montreal workers. Valerie Preston and Joseph C. Cox, "Immigrants and Employment: A Comparison of Montreal and Toronto between 1981 and 1996," *Canadian Journal of Regional Science / Revue Canadienne des Sciences Régionales* 22, nos. 1–2 (1999): 87–111; the data cited are at 94.

9. Preston and Cox, "Immigrants and Employment"; Emil Ooka and Eric Fong, "Globalization and Earnings among Native Born and Immigrant Populations of Racial and Ethnic Groups in Canada," unpublished manuscript, Department of Sociology, University of Toronto, Toronto, 2001; Daniel Hiebert, *The Colour of Work: Labour Market Segmentation in Montreal, Toronto, and Vancouver, 1991,* Research on Immigration and Integration in the Metropolis Working Paper 97-01 (Burnaby, B.C.: Simon Fraser University, 1997); and William J. Coffey, *The Evolution of Canada's Metropolitan Economies* (Montreal: Institute de Recherche en Politiques Publiques / Institute for Research on Public Policy, 1994).

10. Preston and Cox, "Immigrants and Employment," 87–88.

11. Preston and Cox, "Immigrants and Employment," 88–89; Hiebert, "Colour of Work," 22–23.

12. Preston and Cox, "Immigrants and Employment," 92.

13. Hiebert, "Colour of Work," table 2.

14. Victor Piche, Jean Renaud, and Lucie Gingras, "Economic Integration of New Immigrants in the Montreal Labour Market: A Longitudinal Approach," *Population* 57, no. 1 (2002): 57–82.

15. Preston and Cox, "Immigrants and Employment," 92–93.

16. Kazi Stastna, "We're Nicely Educated but Poor," *Montreal Gazette,* February 10, 2004.

17. Stastna, "We're Nicely Educated but Poor."

18. Preston and Cox, "Immigrants and Employment," 92–93.

19. Ooka and Fong, "Globalization and Earnings," 19.

20. François Vaillancourt, *Langue et disparitiés de statut économique au Québec, 1970–1980* (Quebec City: Consiel de la Langue Française, 1988); and François Vaillancourt, *Langue et statut économique au Québec, 1980–1985* (Quebec City: Consiel de la Languge Française, 1991).

21. Vaillancourt, *Langue et status économique au Québec,* 28–30.

22. Dominique Arel, "Political Stability in Multinational Democracies: Comparing Language Dynamics in Brussels, Montreal, and Barcelona," in *Multinational Democracies,* ed. Alain-G. Gagnon and James Tully (Cambridge: Cambridge University Press, 2001), 65–89.

23. Vaillancourt, *Langue et status économique au Québec,* 19–21.

24. P. Villeneuve and D. Rose, "Gender and the Separation of Employment from Home in Metropolitan Montreal, 1971–1981," *Urban Geography* 9 (1988): 155–79; the citation is at 175.

25. Linda Gyulai, "City's Staff Doesn't Reflect Us: Census Reveals Gap; Visible, Cultural Minorities Badly Under-Represented," *Montreal Gazette*, March 26, 2005.

26. McAll, "Breaking Point," 70–77.

27. McAll, "Breaking Point," 77.

28. Diane Coulombe and William L. Roberts, "The French-as-a-Second-Language Learning Experience of Anglophone and Allophone University Students," *Journal of International Migration and Integration / Revue de l'Integration et de la Migration Internationale* 2, no. 4 (2002): 561–79.

29. Jeff Heinrich, "The Immigrant Workers," *Montreal Gazette*, August 2, 2003.

30. Annick Germain and Damaris Rose, *Montréal: The Quest for a Metropolis* (New York: John Wiley & Sons, 2000), 117–18.

31. Heinrich, "Immigrant Workers."

32. Germain and Rose, *Montréal*, 124–25.

33. Germain and Rose, *Montréal*, 136–37.

34. Germain and Rose, *Montréal*, 135–36.

35. Germain and Rose, *Montréal*, 128.

36. Germain and Rose, *Montréal*, 126–27.

37. Germain and Rose, *Montréal*, 154–58.

38. Germain and Rose, *Montréal*, 134–45.

39. Jean-Paul Mbuya Mutombo, "Idenité et performances scolaires: Les élèves issues de groupes minoritaitres au Québec, leur points du vue," *Journal of International Migration and Integration / Revue de l'Intégration et de la Migration Internationale* 4, no.3 (2003): 361–93.

40. Clifford Krauss, "A Sikh Boy's Little Dagger Sets Off a Mighty Din," *New York Times*, June 5, 2002.

41. CBC News Online, "No Kirpans in School, Quebec Court Rules," March 5, 2004; available at http://montreal.cbc.ca/story/canada/national/2004/03/05/Kirpan 040304.html, 3/6/2004.

42. Agence France-Press, "Row over Sikh Boy's Dagger to Go to Canada's Supreme Court," March 6, 2004; available at http://news/yahoo.com/news?tmpl=story2&cid= 1527&u=/afp/20040305/wl_canada_afp/can3/6/2004.

43. CBC Montreal News Online, "Kirpan Case Heads to Supreme Court," October 7, 2004; available at http:// montreal.cbc.ca/regional.servlet/View?filename=qc_ kirpan20041007. Canada Press, "Tobacco Showdown Tops List of Spring Cases for Supreme Court," April 13, 2005.

44. Agence France-Presse, "Montreal Teenager Expelled for Wearing Muslim Veil," September 23, 2003.

45. Jeff Heinrich, "Hijabs a Common Sight at Public Schools, Quebec Human Rights Commission Ruled in 1995 Case," *Montreal Gazette*, September 24, 2003.

46. Robert D. Manning has written most forcefully about this change in such essays as Manning, "Washington, D.C.: The Changing Social Landscape of the International Capital City," in *Origins and Destinies: Immigration, Race and Ethnicity in America*, ed. Silva Pedraza and Ruben G. Rumbaut (New York: Wadsworth, 1996), 373–89; and Manning, "Multicultural Washington, D.C.: The Changing Social and Economic Landscape of a Post-Industrial Metropolis," *Ethnic and Racial Studies* 21, no. 2 (1998): 328–54.

47. For further discussion of this point, see the essays in *Urban Odyssey. A Multicultural History of Washington, D.C.*, ed. Francine Curro Cary (Washington, D.C.: Smithsonian Institution Press, 1996).

48. The continuing economic force of the federal government remains more significant than economic data may suggest. E.g., some analysts have argued that the rapid growth in Hispanic-owned firms, reported later in this chapter, is at least partially a consequence of federal regulations that encouraged government agencies to let contracts to minority-owned businesses. See "A Government Assist: Program Nurtures Minority Firms' Growth," *Washington Post,* June 7, 2004.

49. Bureau of Labor Statistics, U.S. Department of Labor, "State-Area Employment Hours and Earnings Series SAU1100000000001(n)," available at http://data.bls.gov/servlet/SurveyOutputServlet?jrunsessionid=1048691906710189467.

50. Joel Garreau, *Edge City: Life on the New Frontier* (New York: Doubleday, 1991).

51. D'Vera Cohn, "Commuters Crossing Lines: Census Finds Majority Employed Far from Home," *Washington Post,* March 6, 2003.

52. This point is emphasized in the Council of Latino Agencies' 2002 study *The State of Latinos in the District of Columbia;* see Mark Rubin, "Employment," in *The State of Latinos in the District of Columbia,* ed. Council of Latino Agencies / Consejo de Agencias Latinas (Washington, D.C.: Council of Latino Agencies / Consejo de Agencias Latinas, 2002), 121–36).

53. Audrey Singer, Samantha Friedman, Ivan Cheung, and Marie Price, *The World in a Zip Code: Greater Washington, D.C. as a New Region of Immigration* (Washington, D.C.: Brookings Institution Press, 2001), 3.

54. Singer et al., *World in a Zip Code.*

55. Singer et al., *World in a Zip Code.*

56. "Executive Summary," in *State of Latinos in the District of Columbia,* 7.

57. Manning, "Washington, D.C.: Changing Social Landscape," 386.

58. Krissah Williams, "Burgeoning Market Exerts Its Force," *Washington Post,* June 7, 2004.

59. Krissah Williams, "The Changing Face of Arlandia," *Washington Post,* June 7, 2004.

60. Annys Shin, "Broadcasting Company Sets Its Sights High," *Washington Post,* June 7, 2004.

61. Antha Reddy, "All for One and One for All in Quest of Server Efficiency," *Washington Post,* June 7, 2004.

62. "Executive Summary," in *State of Latinos in the District of Columbia,* 22.

63. "Executive Summary," in *State of Latinos in the District of Columbia,* 22.

64. D'Vera Cohn, "One in Five Area Workers Born Abroad," *Washington Post,* February 2, 2004.

65. Cohn, "One in Five Area Workers Born Abroad."

66. Williams, "Burgeoning Market."

67. "Sending Money Back Home," *Washington Post,* June 7, 2004.

68. "Executive Summary," in *State of Latinos in the District of Columbia,* 13.

69. Linda Perlstein, Montgomery Schools at Diversity Landmark," *Washington Post,* October 14, 2003.

70. Kennan Institute Survey of Migrant Families, Kyiv, June–November 2001, questions 67 and 261. For more information about this survey, see the appendix to the present volume.

71. Kennan Institute Survey of Migrant Families, questions 71 and 261. The actual monthly reported income levels appear to be incomplete. However, the hierarchy of earning power reported here is consistent with other data as well as more anecdotal evidence.

72. Kennan Institute Survey of Migrant Families, questions 66 and 67. This experience is consistent with migrant entrepreneurs elsewhere. E.g., Timothy Bates argues that high levels of educational attainment as well as previous employment in the professions before migration helps to explain the relative success of Korean and other migrant merchants in American cities during the 1970s, 1980s, and early 1990s. See Timothy Bates, *Race, Self-Employment, and Upward Mobility: An Illusive American Dream* (Washington and Baltimore: Woodrow Wilson Center Press and Johns Hopkins University Press, 1997).

73. Lily Hyde, "Ukraine's Unwanted Guests," *Kyiv Post*, August 10, 2000.

74. Phan Viet Hung and Fam Van Bang, respectively president and vice president, Vietnamese Association of Ukraine, Kyiv, interview with the author, September 13, 2000.

75. This discussion is based on the findings of the Kennan Institute Survey of Migrant Families conducted in Kyiv between June and November 2001, as well as on an additional Migrant Specialist Survey of Experts and Officials, Kyiv, conducted from March through August 2002. The particular findings reported here are presented in greater detail in Olena Malynovska, Olena Braychevska, Yaroslav Pylynsky, Galina Volosiuk, Nancy E. Popson, and Blair A. Ruble, *Netraditsiinyi immigranti u kyiv* (Washington, D.C.: Kennan Kyiv Project, 2003), section 2.5, prepared by Olena Braychevska and Nancy E. Popson. The Kennan Kyiv Mirgant Survey included seventy-seven school-age children between seven and seventeen years of age, 41.6 percent of whom were born in Kyiv. Approximately one-quarter of the child respondents are entitled to Ukrainian citizenship because they have one parent who is native born to Ukraine.

76. See, e.g., the findings reported in Oleg Shamshur and Tetiana Izhevska, "Multilingual Education as a Factor of Inter-Ethnic Relations: The Case of Ukraine," in *Ethnicity in Eastern Europe: Questions of Migration, Language Rights and Education*, ed. Sue Wright and Helen Kelly (Philadelphia: Multilingiual Matters, 1994), 29–37.

77. Stren and Polese, "Understanding the New Sociocultural Dynamics of Cities," 3.

## Chapter 4—From Saint Jean-Baptiste to Saint Patrick: Montreal's Twisting Path to International Diversity

1. Jay Walz, "Police in Quebec Fight Protesters, with Trudeau in the Stand, Clash Interrupts Parade," *New York Times*, June 25, 1968; Norman Madarasz, "Pierre Bourgault, 1934–2003," http://www.members.1cas.fr/cousture/MADARASZ.htm; Claude Bélingere, "Chronology of Quebec Nationalism, 1960–1991," Marianopolis College Quebec History Web site, http://www2.marianopolis.edu/quebechistory/chronology/national.htm; "Trudeau Timeline," Sympatio NewsExpress, http://www1.sympatico.ca/news/Specials/2000/Trudeau/timeline.htm.

2. E.g., they would see this when he declared his country's War Measures Act to be active in October 1970, during the constitutional debates of 1982, and when he came out of retirement in 1987 and again in 1992 to save his own particular vision of Canada's Charter of Rights.

3. Jay Walz, "Police in Quebec Fight Protesters with Trudeau in the Stand: Clash Interrupts Parade," *New York Times*, June 25, 1968.

4. Madarasz, "Pierre Bourgault."

5. Madarasz, "Pierre Bourgault."

6. "Trudeau Timeline."

7. Lionel Albert, "Cosmopolitan Montrealers Are Abandoning Nationalistic St. Jean-Baptiste Festivities," *Report Newsmagazine,* July 22, 2002; available at http://report.ca/archive/report20020722/p15i0722f.html.

8. Albert, "Cosmopolitan Montrealers."

9. Philippe-Joseph Aubert de Gaspé, *Canadians of Old,* trans. Jane Brierley (Montreal: Véhicule Press, 1996), 114.

10. Jocelyn MacClure, *Quebec Identity: The Challenge of Pluralism,* trans. Peter Feldstein (Montreal: McGill–Queen's University Press, 2003), 150–51.

11. Antonio Barrette, "La grève de l'amiante / The Asbestos Strike," Marianopolis College Quebec History Web site, http://www2.marianopolis.edu/quebechistory/docs/asbestos/6Ad.htm; Claude Bélanger, "Note on Bibliographic Sources to the Study the Asbestos Strike (1949)," Marianopolis College Quebec History Web site, http://www2.marianopolis.edu/quebechistory/docs/asbestos/note.htm; John Gray, "Still a Hero in Asbestos after All These Years," *Globe and Mail* (Toronto), October 2, 2000; Douglas Martin, "After 29 Flamboyant Years, the Mayor of Montreal Is Retiring," *New York Times,* June 28, 1986; Michael T. Kaufman, "Jean Drapeau, 83, Mayor Who Reshaped Montreal" (obituary), *New York Times,* August 19, 1999.

12. MacClure, *Quebec Identity,* 160.

13. Pierre Elliott Trudeau, *La grève de l'amiante: Une étape de la révolution industrielle au Québec* (Montreal: Éditions Cité Libre, 1956).

14. MacClure, *Quebec Identity,* 164–65.

15. Michael Adams, with Amy Landstaff and David Jamieson, *Fire and Ice: The United States, Canada and the Myth of Converging Values* (Toronto: Penguin Canada, 2003), 82.

16. Albert, "Cosmopolitan Montrealers."

17. Elena Cherney, "Pie in William Johnson's Face Leaves Mark on Quebec Holiday," *Ottawa Citizen,* June 25, 1998.

18. Albert, "Cosmopolitan Montrealers."

19. Albert, "Cosmopolitan Montrealers"; Mélanie Brisson, "Une Saint-Jean payante pour les dépanneurs . . . et les pushers!" *Journal de Montréal,* June 25, 2004; Nadia Gaudreau, "Les Québécois on fêté ça!" *Journal de Montréal,* June 25, 2004; and Louise Leduc, "C'est tout? La fête sans défilé déçoit," *La Presse,* June 25, 2004.

20. Matt Stambaugh and Kris Kotarski, "St. Jean-Baptiste: A Quebecois Day Filled with Fun," *Calgary Herald,* July 26, 2003.

21. Lorraine Eisenstat Weinrid, "Trudeau and the Canadian Charter of Rights and Freedoms: A Question of Constitutional Maturation," in *Trudeau's Shadow: The Life and Legacy of Pierre Elliott Trudeau,* ed. Andrew Cohen and J. L. Granatstein (Toronto: Vintage Canada, 1999), 257–82.

22. Pierre Elliott Trudeau, *Federalism and the French Canadians* (Toronto: Macmillan of Canada, 1968), xxi.

23. Trudeau, *Federalism,* 156–57.

24. Trudeau, *Federalism,* 158.

25. Bob Rae, "Trudeau: Hedgehog or Fox?" in *Trudeau's Shadow,* ed. Cohen and Granatstein, 283–94; the quotation is on 293.

26. Walter Stewart, *Shrug: Trudeau in Power* (Toronto: New Press, 1971), 8–10.

27. For an overview of Pearson's considerable accomplishments, see Bruce Thor-

darson, *Lester Pearson: Diplomat and Politician* (New York: Oxford University Press, 1974).

28. Lester B. Pearson, *Mike: The Memoirs of The Right Honorable Lester B. Pearson, Volume 3 (1957–1968)* (Toronto: Signet, 1976), 291–304.

29. Pearson, *Mike*, 254–60; Guy Laforest, *Trudeau and the End of a Canadian Dream*, trans. Paul Leduc Browne and Michelle Weinroth (Montreal: McGill–Queen's University Press, 1995), 68–76.

30. The Trudeau government passed a new law on multiculturalism in 1971 based on the commission's recommendations. Laforest, *Trudeau*, 69.

31. Pearson, *Mike*, 269–70; Bélingere, "Chronology of Quebec Nationalism."

32. Pearson, *Mike*, 233–37.

33. Pearson, *Mike*, 235.

34. Pearson, *Mike*, 237.

35. Bélingere, "Chronology of Quebec Nationalism"; Stewart, *Shrug*, 11–13.

36. Pearson, *Mike*, 358–60.

37. Pearson, *Mike*, 353.

38. Rod McQueen, "Remembering Robert Stanfield: A Good Humoured and Gallant Man," *Policy Options Politiques* 25, no. 2 (February 2004): 8–11.

39. This account is based on Claude Bélingere, "Chronology of the October Crisis 1970, and Its Aftermath," Marianopolis College Quebec History Web site, http://www2 .edu/quebechistory/chronos/October .htm; as well as on "Kidnaped British Official James Richard Cross," *New York Times*, October 7, 1970; Edward Cowan, "Kidnappers of British Aide in Quebec Extend Deadline," *New York Times*, October 8, 1970; and "Quebec Extremists Escalate Separatist Campaign," *New York Times*, October 11, 1978.

40. Graham Fraser, *René Lévesque and the Parti Québécois in Power* (Montreal: McGill–Queen's University Press, 1984), 55–57.

41. "Text of the Canadian Proclamation and Excepts from the Regulations and from Statement by Trudeau," *New York Times*, October 17, 1970; Jay Walz, "Showdown in Canada: Trudeau vs. Quebec Separatists," *New York Times*, October 18, 1970.

42. The principled Douglas declared that the Trudeau government was using "a sledgehammer to crack a peanut." Bèlingere, "Chronology of the October Crisis 1970."

43. Bèlingere, "Chronology of the October Crisis 1970."

44. Edward Cowan, "Canadian Hostage Slain; Body of Laporte Found in Auto Near Montreal," *New York Times*, October 18, 1970.

45. Jay Walz, "Mayor Re-Elected in a Quiet Montreal," *New York Times*, October 26, 1970.

46. Bèlingere, "Chronology of the October Crisis 1970."

47. "Separatists' Earlier Weapons: Bombs, Dynamite, and Robbery," *New York Times*, October 12, 1970.

48. Fraser, *René Lévesque*, 63–70.

49. DeNeen L. Brown, "Separatist Party Loses in Quebec: Nine-Year Hold on Provincial Government Ended by Liberals' Win," *Washington Post*, April 15, 2003.

50. Stewart, *Shrug*, 50–70; J. L. Granatstein, "Changing Positions: Reflections on Pierre Trudeau and the October Crisis," in *Trudeau's Shadow*, ed. Cohen and Granatstein, 295–305; John Saywell, *Agenda 70* (Toronto: University of Toronto Press, 1971); Gérard Pelletier, *La Crise d'Octobre* (Montréal: Editions du Jour, 1971); Robert Bourassa, *Les Années Bourassa* (Montreal: Télémedia, 1977); Claude Ryan, *Le Devoir*

*et la Crise d'Octobre 70* (Montréal: Leméac 1971); and Ron Haggart and Aubrey E. Golden, *Rumours of War* (Toronto: New Press, 1971).

51. Andrew Cohen, "Trudeau's Canada: The Vision and the Visionary," in *Trudeau's Shadow,* ed. Cohen and Granatstein, 307–27; and Laforest, *Trudeau.*

52. Graham Fraser, "October Crisis Still a Black Hole in Our History," *Toronto Star,* October 5, 2003.

53. This discussion of Parezeau's life is based on William Walker, "Parizeau's Own Words His Downfall," *Toronto Star,* November 1, 1995.

54. Sandro Contenta, "Why Montreal Francophones Said 'No,'" *Toronto Star,* November 2, 1995.

55. Daniel Drolet, "By the Numbers, Sovereignty Was Clear Choice of Francophones," *Ottawa Citizen,* November 1, 1995.

56. Robert McKenzie, "PQ Vows to Fight on for Sovereignty," *Toronto Star,* October 31, 1995.

57. The Liberal Party victory over the Parti Québécois in the April 2003 National Assembly elections was seen by some as having closed the issue of Quebec sovereignty for the politically foreseeable future. Brown, "Separatist Party Loses."

58. McKenzie, "PQ Vows to Fight."

59. Rosemary Speirs, "Ill-Advised Remarks Smear Separatist Cause," *Toronto Star,* November 1, 1995.

60. Drolet, "By the Numbers."

61. Brown, "Separatist Party Loses."

62. "List of Quebec Premiers," http://www.nationmaster.com/encyclopedia/List-of-Quebec-premiers; "Premiers of Quebec" (with biographies), http:// canadaonline.about. com/library.bl/blpremque.htm.

63. Bélingere, "Chronology of Quebec Nationalism."

64. Cohen, "Trudeau's Canada"; and Laforest, *Trudeau.*

65. Prime ministers without close ties to Quebec during this period include the short-lived governments of Joe Clark (1979–80), John Turner (1984), and Kim Campbell (1993). "Prime Ministers of Canada—1867 to Date," http://www.parl.gc.ca/information/about/people/key/PrimeMinster.asp?Language-E.

66. "The Great Names of the French Canadian Community: Jean Drapeau," http://edimage.ca/edimage/grandespersonnages/en.carte_r02html; Martin, "After 29 Flamboyant Years;" Kaufman, "Jean Drapeau."

67. Henri Lustiger-Thaler and Eric Shragge, "The New Urban Left: Parties without Actors," *International Journal of Urban and Regional Research* 22, no. 2 (1998): 233–44; the citation here is at 235–37.

68. Lustiger-Thaler and Shragge, "New Urban Left."

69. Jean-François Léonard and Jacques Léveillée, *Montréal after Drapeau* (Montréal: Black Rose Books, 1986), 129.

70. Léonard and Léveillée, *Montréal,* 133.

71. Lustiger-Thaler and Shragge, "New Urban Left," 237.

72. Lustiger-Thaler and Shragge, "New Urban Left," 237; Michelle LaLonde, "I'll Do Better Job This Time, Dore," *Montreal Gazette,* April 28, 1998; and Lina Gyulai, "One Year after Mayoral Race, Back in Business," *Montreal Gazette,* October 28, 1999.

73. Lustiger-Thaler and Shragge, "New Urban Left," 238–40.

74. Pierre Bourque, "Homage to Jean Drapeau," http://pages.globetrotter.net/drapeau/bourque-end.html.

75. Bourque's identification with his city, and with its cultural diversity, may be seen in his memoirs: Pierre Bourque, *Ma passion pour Montréal* (Montréal: Reader's Digest Association (Canada), 2002).

76. This discussion of Bourque's career is based on William Johnson, "Pierre Bourque: He's Back," *Globe and Mail,* February 11, 2003.

77. Charlie Fidelman, "Deep Pride in His Roots," *Montreal Gazette,* June 23, 2003.

78. Bourque, *Ma passion pour Montréal.*

79. "P. Bourque.bio," http://www.pierrebourque.org/biographie.html.

80. McKenzie, "PQ Vows to Fight"; Drolet, "By the Numbers"; and Contenta, "Why Montreal Francophones said 'No.'"

81. E.g., see his message to the "Inclusion by Design" conference in 2001: "Message du Maire de Montréal M. Pierre Bourque," http://www.inclusionbydesign.com/worldcongress.bourquef.htm.

82. "Pierre Bourque," *L'Encylopédie de L'Agora,* 2004, http://agora.qc.ca/mot.nsf/Dossiers/Pierre-Bourque.

83. Michelle LaLonde, "Quebec Denies Mayor's Claim of Highway Work," *Montreal Gazette,* July 23, 1998; "Montrealers Deserve a Voice," *Montreal Gazette,* July 19, 1999.

84. Peter Black, "Montreal Is Now a City Forever Changed," *Press Republican Online,* November 9, 2001, http://www.pressrepublican.com/Archive/2001/11_2001/11092001pb.htm.

85. Black, "Montreal."

86. E.g., see one typical news item of the period: Charlie Fidelman and Darren Becker, "Huge No to Megacities," *Montreal Gazette,* December 11, 2000.

87. Black, "Montreal."

88. "Gérald Tremblay: First Mayor of the New City of Montréal / Gérald Tremblay: Premier maire de la nouvelle ville de Montréal," http:www1.cata.ca/townhall/mayors/grtemblay.cfm.

89. Black, "Montreal."

90. Brown, "Separatist Party Loses."

91. Johnson, "Pierre Bourque."

92. Johnson, "Pierre Bourque."

93. Johnson, "Pierre Bourque."

94. Jeff Heinrich, "Borque's Policy Hits 'Cultural' Bump: Vision Montreal Party Struggles to Find the Right Term to Define Openness to Minorities," *Montreal Gazette,* October 3, 2003.

95. Peter Black, "Pierre Bourque to the ADQ," *Log Cabin Chronicles,* March 2004, http://www.tomifobia.com/black/pierre_bourque.shtml.

96. "Bourque Returns to City Hall," *Montreal Gazette,* April 22, 2003.

97. Glenn Wanamaker, "One Island, One City, Another PQ Test," CNews, November 1, 2001, http://www,canoe.ca/CNEWSPolitics0111/01_wanamaker-can.html; Philip Authier and Sean Gordon, "Demerger foe pursues Charest," *Montreal Gazette,* March 30, 2003; Jean Corriveau, "Project de loi sur les défusions—Les maires des grandes villes rencontrent Charest," *Le Devoir,* June 26, 2003; and François Cardinal, L'administration Tremblay dépense de plus en plus pour combattre la défusion," *La Presse,* July 31, 2003.

98. Andrew Sancton, "Beyond the Municipal: Governance for Canadian Cities," *Political Options Politiques* 25, no. 2 (2004): 26–31; the quotation is on 29–30.

99. Robert Dutrisac, "Avant les défusons, l'adhésion," *Le Devoir,* June 8, 2003; Robert Dutrisac, "Les villes de l'ouest de L'île pourront se regrouper," *Le Devoir,* June 19, 2003; Presse canadienne, "Le plan Tremblay pour èviter les dèfusions," August 15, 2003; "Tremblay Offers Megacity Compromise," *Canada.conNews,* August 15, 2003; Kevin Dougherty, "Liberals Back Tremblay: Charest Likes Decentralization Proposals," *Montreal Gazette,* August 17, 2003; and Jeanne Corriveau, Le plan de décentralisation du maire Tremblay inquiète les syndicats," *Le Devoir,* August 19, 2003.

100. Sancton, "Beyond the Municipal," 30.

101. Peter Black, "Shooting for Mega-City Mayor," *Log Cabin Chronicles,* January 3, 2001; Gerald Tremblay, "Citizens of Canada Unite," *Globe and Mail,* January 11, 2002; Bindu Suresh, "Mayor Unfazed by Demerger Movement," *Montreal Gazette,* August 24, 2003; and Black, "Montreal."

102. Konrad Yakubski, "Montreal Faces Wrath of Urban Separatists," *Globe and Mail,* June 16, 2004.

103. Linda Gyulai, "Tremblay Clashes with Party Dissident, 'Think about Political Future,'" *Montreal Gazette,* August 19, 2003; and CBC, "Mayor Faces More Dissent over Demerger," http://montreal.cbc.ca/regional/servlet/View? filename=qc_mayor 20032108.

104. Sébastien Rodrique, "Défusionnistes contra Fusionistes: Bourque veut des élections municipales référendaires," *La Presse,* August 26, 2003; Linda Gyulai, "Delay Demerger Vote Until 2005: Bourque," *Montreal Gazette,* August 26, 2003; CBC, "Few Backers for Bourque-Style Reform," August 26, 2003, http://montreal.cbc .ca/regional/servlet/PrintStory?filename-qc_bourque20032608&region; Elisabeth Kalbfuss, "Mayor Asks Two Years to Prove City Can Work," *Montreal Gazette,* August 28, 2003; Gyulai, "Megacity Meetings Scheduled," *Montreal Gazette,* August 20, 2003; and Gyulai, "Tremblay Begins Boroughs Meetings," *Montreal Gazette,* September 3, 2003.

105. Graeme Hamilton, "Anglos Threaten Montreal Mega-City, Pundits Warn," *National Post,* October 9, 2003.

106. Irwin Block, "More Suburb-Dwellers Want to Hold Referendums," *Montreal Gazette,* September 5, 2003; and Allison Hanes, "Mayors Warn They'll Fight Dismantling of Their Cities," *Montreal Gazette,* September 9, 2003.

107. LCN, "Défusions: Anglophones et francophones ne sont pas sur la même longueur d'onde," http://www2.canoe.com/infos/regional/archives/2003/09/20030904-173045.html; Presse Canadienne, "'Défusions': Pas de moratoire sur les référendums," http://www2.canoe.com/infos/natinal/archives/2003/09/20030905-165328.html; and Radio Canada, "Une majorité de Montréalais s'opposent aux défusions municipales," http://radio-canada.ca/nouvelles/Index/nouvelles/200309/04/019-sondage-defusins .shtml.

108. Kevin Dougherty, "Megacity Will Survive Borough Vote: Minister," *Montreal Gazette,* October 8, 2003; François Cardinal, "*La Presse* met la main sur le nouveau plan d'urbanisme de Montréal: D'abord du transport et un accès aux berges," *La Presse,* October 8, 2003; Linda Gyulai, "Who Ordered Demerger Probe of Sûreté?" *Montreal Gazette,* October 11, 2003; Sébastien Rodrigue, "Défusionistes interrogés: La SQ blâmée," *La Presse,* October 11, 2003; and Rodrigue, "La SQ enquête sur les défusionnistes," *La Presse,* October 10, 2003.

109. Martin Ouellet, "Relations francophones–anglophones 'Le défusions seront néfastes,' selon le PQ," *Presse Canadienne,* May 17, 2004.

110. Discussion and direct comments reflecting public perceptions of the issue may be found on the "Carnet web de Laurent Gloaguen, *Tout n'est que vanité*," Web site, http://embruns.net/quebec/001176.html.

111. Jeanne Corriveau, "Le maire Gérald Tremblay met en garde les défusionnistes: Les ville reconstituées géreront seulement 30 percent de leur budget," *Le Devoir,* October 15, 2003; Robert Dutrisac, "Modification au project de loi 9: Québec envisage d'imposer un modèle uniforme aux villes défusionnées," *Le Devoir,* October 17, 2003; Yves Bellavance, "Le véritable enjeu des défusions, le refus de partager les richesses," *Le Devoir,* November 10, 2003; and Miro Cernetig, "Undoing the Montreal Megacity," *Toronto Star,* December 6, 2003.

112. Unnati Gandhi, "Advance Poll Lineups Delight Demergerites," *Montreal Gazette,* June 14, 2004.

113. François Cardinal, "'Désinformation,' 'démagogie': Les défusionnistes accusés de mentir," *La Presse,* May 15, 2004; Marie-Andrée Chouinard, "En route vers les référendums dur la défusion," *Le Devoir,* May 17, 2004.

114. "Le vote des défusions marqué par une forte participation," *Canoë Infos,* June 14, 2004, http://www2.infini.com.

115. Bernard Descôteaux, "Un project imparfait," *Le Devoir,* November 29–30, 2003; and Denis Lessard, "Québec propose une 'CUM améliorée,' Montréal morcelé?" *La Presse,* November 29, 2003.

116. Henry Aubin, "Montreal's Man of the Year: Mayor Gérald Tremblay Was a star on the Municipal Scene in 2003," *Montreal Gazette,* December 28, 2003; and "Mayor Rewards New Allies," *Montreal Gazette,* January 27, 2004.

117. "Le parti de Gérald Tremblay aurait usé de fonds publics contre les défusions," *Canoë Infos,* June 11, 2004, http://www2.infini.com.

118. Dida Berku, "Megacity Has Made Great Progress in Two Years, Dynamic and Modern: Agglomeration Council Is Not Your Father's MUC," *Canada.comNews,* http://www.canada.com; Marc Lamarche, Nancy Neamtan, Marcel Pedneault, and Pierre Richard, "Construire ensemble notre quartier et notre ville," *Le Devoir,* June 18, 2004; and Miro Cernetig, "Montrealers Get Chance to Reverse Mergers," *Toronto Star,* June 19, 2004.

119. François Cardinal, "Défusions après le OUI viendra une pluie de poursuites," *La Presse,* May 18, 2004; Sébastien Rodrigue and François Cardinal, "Défusions: Un cas de fraude à Saint-Laurent," *La Presse,* May 18, 2004; CBC Montreal, "Charest Tones Down Demerger Hopes," May 18, 2004, http://montreal.cbc.ca/regional/servlet/View?filename=qc_charest20040518.

120. Pierre Bourque, "Libre opinion: 'Une île, une ville'—Un project de societé pour les Montréalais," *Le Devoir,* May 23, 2004.

121. Linda Gyulai, "Mayor Plays the Suburb Card," *Montreal Gazette,* January 26, 2004.

122. "Mergers Beefed Up Fire Departments," *Montreal Gazette,* June 15, 2004.

123. Allison Lampert, "Poverty Changing Address: Report's Map Shows Poor Have Spread Out Across the Island," *Montreal Gazette,* January 16, 2004.

124. CBC Montreal, "Défusion versus démembrement," May 23, 2004, http://www.montreal.cbc.ca/backgrounder/dermerger; "Sign Those Registers," *Montreal Gazette,* May 19, 2004; Antoine Robitalle, "Défusions: Ce n'est qu'un début—Les référendums gagnés, les défusionnistes s'attaqueront à la loi 9," *Le Devoir,* May 18, 2004; Brena Branswell, "Half of Montreal Appears Headed for Demerger Votes," *Mon-*

*treal Gazette,* May 19, 2004; CBC Montreal, "Pro-Merger Forces Promise to Speak Up," May 19, 2004, http://montreal.cbc.ca/regional/servlet/PrintStory ?filename=qc_merger 20040519&region=; Antoine Robitalle, Guillaume Bourgault-Côté, "Raz-de-marée dé-fusionniste: 22 référendums sur l'île de Montréal—Une pluie de consultations ailleurs au Québec," *Le Devoir,* May 21, 2004; "Les défusionnistes gagnent la première manche!" *Canoë Infos,* May 21, 2004, http://www2.infini.com; CBC Montreal, "Mayor Stunned by Support for Demergers," May 21, 2004, http://montreal.cbc.ca/ regional/servlet/View?filename=qc_demerger20040521; and Robert Dutrisac and An-toine Robitalle, "Charest votera NON le 20 juin: Un 'love-in' contre les défusions à Westmount la veille des réféndums," *Le Devoir,* June 1, 2004.

125. "Demerger Campaign Turns Ugly in Quebec City," *Global Quebec News,* June 15, 2004, http://www.canada.com/montreal/globaltv/index.htm1, 6/15/2004.

126. LCN, "Débat animé entre Gérald Tremblay et Peter Trent," June 7, 2004, http://www2.canoe. com; CBC Montreal, "Tone Raised during Demerger Debate," June 7, 2004, http://montreal.cbc.ca/reginal/servlet/View?filename=qc_debate20040607; CBC Montreal, "Montrealers Had Their Own Debate Monday," June 15, 2004, http://montreal.cbc.ca/regional/servlet/View?filename=qc_debate20040615; Radio-Canada, "Défusions: Gérald Tremblay Pourfend Peter Trent," June 15, 2004, http://www.radio-canada.ca/nouvelles/Index/nouvelles/200406/14/018-Defusions-Debat.shtml.

127. "Megacity's Number Might Be Up. Trent vs. Tremblay Reprise," *Montreal Gazette,* June 15, 2004.

128. Unnati Gandhi, "Advance Turnout Looks Good for Yes," *Montreal Gazette,* June 15, 2004; Norman Delisle, "Le PQ ne touchera pas aux défusions municipales," *Canoë Infos,* June 17, 2004, http://www2.infinit.com/cgi-bin/imprimer.cgi?id=143120; and Radio-Canada, "Le camp du NON est revigoré," June 18, 2004, http://www.radio-canada.ca/regions/Montreal/nouvelles/200406/18/004-zampino-bosse.shtml.

129. François Cardinal, "Référendums sur les défusions: Tremblay concède la vic-toire dans 11 villes," *La Presse,* June 17, 2004; Radio-Canada, "Le maire de Montréal ne concède rien," June 17, 2004, http://radio-canada.ca/regions/montreal/version_ imprimable.asp?nv=/regions/Montreal/. These two stories present diametrically opposed characterizations of the same remarks by Tremblay a few days before the election, demonstrating the confusion that reigned in the city at the time.

130. These newspapers included both the English-language *Montreal Gazette* and the French-language *La Presse, Le Devoir,* and *Le Soleil* in Montreal, as well as Ottawa's French-language daily *Le Droit.* Radio-Canada, "*The Gazette* se prononce contre les défusions," June 18, 2004, http://www.radio-canada.ca/regions/special/nouvelles/ Defusions_Montreal/200406/18/001-GAZETTE-DEFUSION.shtml.

131. LCN, "Jour de référendums: 89 municipalités jouent leur avenir," June 20, 2004, http://www2.canoe.com/cgi-bin/imprimer.cgi?id=143282.

132. Linda Gyulai, "Demergers Alter Quebec's Urban Landscape," *Montreal Gazette,* June 21, 2004.

133. Jonathan Card, "Vote on Megacities," *Globe and Mail,* June 21, 2004.

134. Linda Gyulai, "Peter Trent to Opt Out of Westmount Elections: 'My Job is Done'; Fierce Merger Foe Says He Hasn't Spoken to Provincial or Federal Parties About Running," *Montreal Gazette,* February 1, 2005.

135. Nuri Cernetig, "Quebec's Megacity Dream Dies," *Toronto Star,* June 21, 2004; Robert Dutrisac, "Défusions: Fournier prédit l'apaisement des tensions linguistiques à

Montréal," *Le Devoir,* June 22, 2004; Denis Lessard, "Charest estime que Montréal sont gagnante de l'exercice référendaire," *La Presse,* June 23, 2004.

136. Cernetig, "Quebec's Megacity Dream Dies;" Robert Dutrisac, "Charest nie tout clivage linguistique," *Le Devoir,* June 23, 2004; Lessard, "Charest estime que Montréal sort gagnante de l'exercise référendaire"; Dutrisac, "Défusions"; Kevin Dougherty, "Vote Not About Language," *Montreal Gazette,* June 23, 2004. For a discussion of the peculiarities of Montréal-Est, see Linda Gyulai, "Anchored by Deep Roots: Montreal East Can Cause Outsiders to Scratch Their Heads—Why Would This Tiny Enclave Want to Separate from Montreal? For Those Who Know Its History, There's No Secret," *Montreal Gazette,* July 22, 2004.

137. Cernetic, "Quebec's Megacity Dream Dies."

138. "A Rude Awakening," *Montreal Gazette,* June 18, 2004.

139. This point was emphasized by the doyen of Montreal urban analysts, Henry Aubin of the *Montreal Gazette.* Aubin, "Home Ownership Key in Votes," *Montreal Gazette,* June 25, 2004.

140. Cernetic, "Quebec's Megacity Dream Dies"; André Beauvais, "Montréal déchiré," *Journal de Montréal,* June 21, 2004.

141. LCN, "Montréal s'agrandit, Gérald Tremblay se réjouit," June 20, 2004, http://www2. canoe.com/cgi-bin/imprimer.cgi?id=143335; Dougherty, "Vote Not About Language."

142. Dougherty, "Vote Not About Language."

143. There was little reason to believe that city negotiators were going to be forgiving when they met with the various commissions established to negotiate fifteen municipal divorces to take place by the end of 2005. Dougherty, "Vote Not About Language"; Linda Gyulai, "Tremblay Reassesses His Team," *Montreal Gazette,* June 22, 2004.

144. LCN, "Les défusions coûteront prés de 15 M$," July 10, 2004, http://www2 .canoe.com/cgi-bin/imprimer.cgi?id=144883. CBC Montreal, "Plus de poids aux villes defusionnees?" March 9, 2005.

145. André Beauvais, "La défusion est devenue un casse-tête colossal," *Journal de Montréal,* October 8, 2004.

146. Radio Canada, Montréal "Les col bleus de Montréal ont une convention," October 5, 2004, http://radio-canada.ca/regions/montreal/version_imprimable.asp?nv=/ regions/Montreal/, 10/5/2004.

147. Radio Canada, Montréal, "Manque à gagner de 35 millions à Montréal," October 26, 2004, http://radio-canada.ca/regions/montreal/version_imprimable.asp?nv=/ regions/Montreal/, 10/27/2004; Caroline Touzin, "Rapport sur la réforme électorale Montréal devra sortir son portefeuille," *La Presse,* October 27, 2004.

148. Sébastien Rodrigue, "Les Montréalais prêts à réélire Tremblay," *La Presse,* October 27, 2004; Radio Canada, Montréal, "Gérald Tremblay favori des Montréalais," October 27, 2004, http://radio-canada.ca/nouvelles/print.asp?nv=noulles/Index/ nouvelles/200410/27/01, 10/27/2004; Rodrique, "L'obsession du maire Tremblay: Les finaces," *La Presse,* October 28, 2004; Rodrique, "Réforme du mode électoral: Le débat s'annonce houleux á Montréal," *La Presse,* November 23, 2004.

149. CBC Montreal, "Budget Brings Tax Cuts For Most," November 26, 2004, http://montreal.cbc.ca/regional/servlet/PrintStory?filename=qc-budget20041125 &region = Montreal, 11/26/2004; "Baisse de taxes pur une majorité Budget de Montréal à 3,98 milliards en 2005," *Presse Canadienne,* November 25, 2004.

150. CBC Montreal, "Budget Brings Tax Cuts For Most," November 26, 2004, http://montreal.cbc.ca/regional/servlet/PrintStory?filename=qc-budget20041125.

151. Karim Benessaieh, "Les élus des villes défusionnées 'frustrés' par Québec: Le conseil d'agglomération jugé 'invivable,'" *La Presse,* November 18, 2004.

152. Émilie Côté, "Project Montréal: Un nouveau parti municipal," *La Presse,* November 18, 2004.

153. CBC Montreal, "Bourque to Run as Mayor Again," February 24, 2005; LCN, "Pierre Bourque pret pour l'election," February 24, 2005; Linda Gyulai, "Montreal Politics Dying, Some Fear. Citywide Parties Might Be History—Boroughs Become the Seat of Power as Municipalities' Autonomy Expands," *Montreal Gazette,* February 29, 2005.

154. "Saint-Patrick: C'est le grand défilé," *La Presse,* March 14, 2004; Office of the Prime Minister (Ottawa), "Notice: Prime Minister Paul Martin Travels to Montreal," March 11, 2004, http://pm.gc.ca/eng/news.asp?id=113.

155. Pearson, *Mike,* 353.

156. Brown, "Separatist Party Loses."

157. "We're All Irish," *Montreal Gazette,* March 16, 2002.

158. Peter Black, "A Legacy of St. Patrick in Montreal," *Press Republican Online,* March 16, 2001, http://www.pressrepublican.com/Archive.

159. Black, "Legacy of St. Patrick."

160. Albert, "Cosmopolitan Montrealers."

161. Rene Bruemmer, "Irish Eyes to Smile on First Black Parade Queen," *Montreal Gazette,* February 9, 2004; available at http://www.athletics.mcgill.ca.varsity_sports _article.ch2?aticle_id=225.

162. Bruemmer, "Irish Eyes to Smile."

163. Lorraine Carpenter, "The Green Party: Where to Kiss It Up and Piss It Up This St. Patrick's Day," *Montreal Mirror,* March 2004, http://www.montrealmirror.com/ meat/music3/html.

## Chapter 5—Regime Change in Washington

1. Michael Powell and Vanessa Williams, "D.C. Council Chairman David A. Clarke Dies; Political Career Began in Civil Rights Era," *Washington Post,* March 29, 1997.

2. Hamil R. Harris and Yolanda Woodlee, "Friends, Colleagues Bid Clarke Goodbye; D.C. Council Chairman 'Didn't See Color; He Saw People as People,'" *Washington Post,* April 4, 1997; Vincent S. Morris, "Hundreds Pay Last Respects to Clarke," *Washington Times,* April 4, 1997.

3. David L. Lewis, *District of Columbia: A Bicentennial History* (New York: W. W. Norton, 1976), 70–71; Howard Gillette Jr., *Between Justice and Beauty: Race, Planning, and the Failure of Urban Policy in Washington, D.C.* (Baltimore: Johns Hopkins University Press, 1995), 49–68.

4. Harris and Woodlee, "Friends, Colleagues Bid Clarke Goodbye"; Morris, "Hundreds Pay Last Respects to Clarke."

5. Powell and Williams, "D.C. Council Chairman David A. Clarke Dies."

6. Powell and Williams, "D.C. Council Chairman David A. Clarke Dies."

7. Powell and Williams, "D.C. Council Chairman David A. Clarke Dies."

8. Powell and Williams, "D.C. Council Chairman David A. Clarke Dies."

9. Marion Barry Jr., "'A Giant for This City'; for Dave Clarke, Washington Was a Lifelong Home and a Lifelong Cause," *Washington Post,* March 29, 1997.

10. Loose Lips, "Moving beyond Clarke," *Washington City Paper,* February 3–13, 1997.

11. Barry, "'Giant for This City.'"

12. Lewis, *District of Columbia,* 181–82. Julius Hobson Sr. and his family are featured prominently in Burt Solomon's *The Washington Century: Three Families and the Shaping of the Nation's Capital* (New York: HarperCollins, 2004). Solomon traces the Hobsons, together with the family of Jewish real estate magnate Morris Cafritz, and the doyens of Louisiana politics and transplanted Washingtonians, members of the Lindsay Boggs dynasty, throughout the twentieth century as a way of telling the story of Washington's emergence from the proverbial sleepy Southern town to a national metropolis. The story of Julius Hobson's 1964 Rat Relocation Rallies may be found on pages 154–56.

13. Harry S. Jaffe and Tom Sherwood, *Dream City: Race, Power, and the Decline of Washington, D.C.* (New York: Simon & Schuster, 1994), 44–66.

14. Jaffe and Sherwood, *Dream City,* 94.

15. One of the more concise accounts of the 1968 riots and their impact on local Washington politics is to be found in Jaffe and Sherwood, *Dream City,* 67–87.

16. Jaffe and Sherwood, *Dream City,* 82.

17. Jaffe and Sherwood, *Dream City,* 30–51.

18. Jaffe and Sherwood, *Dream City,* 99–105.

19. Jaffe and Sherwood, *Dream City,* 103.

20. Jaffe and Sherwood, *Dream City,* 104.

21. As quoted in Jaffe and Sherwood, *Dream City,* 93.

22. Jonetta Rose Barras, *The Last of the Black Emperors: The Hollow Comeback of Marion Barry in the New Age of Black Leaders* (Baltimore: Bancroft Press, 1998), 179–208.

23. Loose Lips, "Moving beyond Clarke."

24. Howard Gillette Jr., "Washington, D.C., in White and Black," in *Composing Urban History and the Constitution of Civic Identities,* ed. John J. Czaplicka, Blair A. Ruble, and Lauren Crabtree (Washington and Baltimore: Woodrow Wilson Center Press and Johns Hopkins University Press, 2003), 192–210.

25. Ruben Castaneda, "The Man in the Eye of the Storm: Immigrant Shot in Mt. Pleasant Called Hard-Working, Shy," *Washington Post,* May 13, 1991.

26. This account based on Gabriel Escobar and Nancy Lewis, "For Every Witness, Yet Another Account of the 'Real' Story," *Washington Post,* May 7, 1991.

27. Nancy Lewis and James Rupert, "D.C. Neighborhood Erupts after Officer Shoots Suspect: Crowd of Hundreds Confronts Police, Sets Cruisers Ablaze in Mt. Pleasant," *Washington Post,* May 6, 1991; Carlos Sanchez and Rene Zanchez, "Dixon Imposes Curfew on Mt. Pleasant Area after Clash for a Second Night: Skirmishes, Looting Spread under Cloud of Tear Gas," *Washington Post,* May 7, 1991; "Curfew Leaves Mount Pleasant Area Quieter: Sporadic Incidents Reported on 3rd Night," *Washington Post,* May 8, 1991; Gary Fields, "Dixon Declares Curfew: Rioters Rampage Again in NW," *Washington Times,* May 8, 1991; Rene Sanchez, "3rd Night of Curfew Quiet but

Uneasy in Mt. Pleasant Area," *Washington Post,* May 9, 1991; and Keith Harriston, "Life Gets Back on Track in Mt. Pleasant: Dixon Lifts Curfew but Police Maintain Strong Presence in Area," *Washington Post,* May 10, 1991.

28. Keith Harrison, "Life Gets Back on Track in Mt. Pleasant: Dixon Lifts Curfew but Police Maintain Strong Presence in Area," *Washington Post,* May 10, 1991.

29. Overall, the population in census tracts clustered in this area would increase during the 1990s, even as Washington's population as a whole would decline. Median household income in some census tracts within the area rose by as much as 50 percent. Meanwhile, the African-American population fell sharply even as the Latino and non-Hispanic white populations grew. These trends, which were just beginning to become felt as Mount Pleasant exploded, were discussed a dozen years later in Debbi Wilgoren, "A Core of City, Old-Timers, Recent Arrivals Seek a New Balance. Population Shifts Leave Some Poorer Residents Worried," *Washington Post,* July 10, 2003.

30. Ruben Castaneda and Nell Henderson, "Simmering Tension Between Police, Hispanics Fed Clash," *Washington Post,* May 6, 1991.

31. Carlos Sanches and Nell Henderson, "D.C. Hispanics Find Voice, but Power Is Elusive," *Washington Post,* (May 11, 1991).

32. This account is based on Olivia Cadaval, "Tirarlo a la calle / Taking It to the Street: The Latino Festival and the Making of Community," *Washington History* 4, no. 2 (1992–93): 40–55. Cadaval's article is accompanied by numerous photographs by Rick Reinhard.

33. Cadaval, "Tirarlo a la Calle."

34. Cadaval, "Tirarlo a la Calle."

35. Cadaval, "Tirarlo a la Calle."

36. Cadaval, "Tirarlo a la Calle."

37. These events are recounted in Jonetta Rose Barras's lively account of the Barry "comeback," *Last of the Black Emperors.*

38. Gillette, "Washington."

39. Barras, *Last of the Black Emperors,* 267–69.

40. Gillette, "Washington."

41. Krisha Roy, "Socio-Demographic Profile," in *The State of Latinos in the District of Columbia,* ed. Council of Latino Agencies / Consejo de Agencias Latinas (Washington, D.C.: Council of Latino Agencies / Consejo de Agencias Latinas, 2002), 1–30, esp. fig. 1.1 on 5; available at http://www.consejo.org.

42. Barras, *Last of the Black Emperors,* 180–82.

43. U.S. General Accounting Office, *Report of the Ranking Minority Member, Subcommittee on the District of Columbia, Committee on Government Reform, House of Representatives: District of Columbia: Reporting Requirements Enacting by Congress (GAO-01-1134, September 2001)* (Washington, D.C.: U.S. General Accounting Office, 2001).

44. Bill Miller, "In Her Court, an Open-and-Shut Case: Christian Shows D.C. Schools Litigants That She's Serious about Safety," *Washington Post,* August 24, 1997. Loose Lips, "Gen. Becton's Waterloo," *Washington City Paper,* August 29–September 4, 1997; available at http://www.washington citypaper.com/archives/lips/lips0829.html. Maria Tukeva, "The Damage That Can't Be Repaired: When My D.C. School Was Closed, Windows of Opportunity Shut, Too," *Washington Post,* November 9, 1997. Debbi Wilgoren, Valerie Strauss, and David A. Vise, "One Year Later, Becton Still Struggles," *Washington Post,* November 18, 1997. "Q&A with Julius Becton Jr.," *Washington Post,* April 14, 1998; available at http://discuss.washingtonpost.com/wp-srv/zforum/levey/ bob0414.htm.

45. Consumer and Regulatory Affairs, District of Columbia Government, "News Release: City Council Votes on Master Business License Repeal Act; Master Business License Deadline Extended," June 11, 2003; available at http:dcra.dc.gov/newsroom/2003/June/06_11_03.htm. Consumer and Regulatory Affairs, District of Columbia Government, "News Release: Master Business License Program Modified to Basic Business Licensing," July 8, 2003; available at http:dcra.dc.gov/newsroom/2003/July_7_08_03.htm.

46. Unless otherwise indicated, this account is based on Barras, *Last of the Black Emperors,* 185–208.

47. D.C. Watch, "Board of Elections Challenge to Anthony Williams's Nominating Petitions, 2002 Election," available at http:// www.dcwatch.com/archives/election 2002/williams00. htm.

48. Unless otherwise indicated, this account is based on Barras, *Last of the Black Emperors,* 203–8.

49. "Mayor Barry Announces His Decision Not to Run for Public Office: 'My Service Is a Personal Testament of Triumph,' says Mayor Barry," D.C. Office of Communications, Washington, May 21, 1998; available at http://www.dcwatch.com/mayor/980521b.htm.

50. "Statement of Former D.C. Mayor Marion Barry Jr.," *Washington Post Newsweek Interactive,* April 4, 2002; available at http://www.washingtonpost.com/wp-dyn/articles/A60771-202Apr4.htm.

51. Yolanda Woodlee and Lori Montgomery, "Barry Declines to Confirm Bid but Offers Hints," *Washington Post,* June 3, 2004.

52. Lori Montgomery, "Barry Launches Campaign for D.C. Council: Former Mayor Rails Against City Services, 'Failing' Schools," *Washington Post,* June 13, 2004.

53. Elissa Silverman, "Loose Lips: Checks and Balances," *Washington City Paper,* July 2–8, 2004; Montel Reel, "Ex-Campaign Manager Says Barry Isn't Up to Job," *Washington Post,* July 3, 1004; Yolanda Woodlee and Montel Reed, "Amid Rumors, Queries, Barry Soldiers On," *Washington Post,* August 8, 2004; Yolanda Woodlee, "Barry's Finance Report Is Incomplete," *Washington Post,* August 26, 2004.

54. Montgomery, "Barry Launches Campaign for D.C. Council."

55. Paul Schwartzman, "In a Changed Ward, Barry Is Back," *Washington Post,* December 27, 2004.

56. Lori Montgomery and Yolanda Woodlee, "Voters in D.C. Say Yes to Barry—Again," *Washington Post,* September 15, 2004; Paul Schwartzman and Hamil R. Harris, "Remarkable Victory Feels So Familiar," *Washington Post,* September 15, 2004; David Nakamura, "East of Anacostia, Anger Is Power," *Washington Post,* September 15, 2004); David Nakamura and Yolanda Woodlee, "In D.C.'s Eastern Wards, Discontent Was Clear," *Washington Post,* September 16, 2004; Matthew Cella, "Barry Back on Council after Ward 8 Landslide," *Washington Times,* November 3, 2004.

57. Montgomery and Woodlee, "Voters in D.C. Say Yes to Barry—Again."

58. "Vote Kwame Brown," at http://www.votekwamebrown.org/kwame/Index.nsf/ViewInfo/ biography?OpenDocument.

59. "Leadership," at http://www.dcdemocrats.org/ward7/htmml.

60. Official biographies of Council members may be found on the D.C. Council's Web site, http://www.grc.dc.gov/city_council/cwp/view.asp?a=3&q=447541.

61. D'Vera Cohn, "For Catania, Politics Goes Beyond Gay; Activists' Reaction Mixed on GOP Council Member," *Washington Post,* December 11, 1997. Ward-level

election results may be found on the D.C. Board of Elections and Ethics Web site, http://www.dcboee.org/htmldocs/ncche.htm.

62. Vanessa Williams, "Catania to Become Independent Today," *Washington Post,* September 29, 2004.

63. District of Columbia Board of Elections and Ethics, "Certified Summary Results, Presidential, November 2, 2004," available at http://www.dcboee.org/information/elec_2004pres_general_2004_results .shtm.

64. "The 1998 Election," *Washington Post,* November 5, 1998.

65. Carol Schwartz's official biography may be found on the D.C. Government's City Council Web site, at http://www.dccouncil.washington.dc.us/SCHWARTZ/bio.htm.

66. District of Columbia Board of Elections and Ethics, "Final and Complete Election Results for the November 5, 1998 General Election," available at http://www .dcboee.org/htmldocs/CNDSUM.LST.

67. District of Columbia Board of Elections and Ethics, "Final and Complete Election Results for the September 15, 1998, Primary Election," available at http://www .dcboee.org/htmldocs/0910resu.htm.

68. In the spirit of full disclosure, the author is a member of this organization.

69. Vanessa Williams, "After 15 Years, Smith Faces Tough Race for Ward 1 Seat," *Washington Post,* July 19, 1998.

70. Williams, "After 15 Years, Smith Faces Tough Race"; Jim Graham's official biography may be found on the D.C. Government's City Council Web site, http://www.grahamwone.com/biography.index.html.

71. District of Columbia Board of Elections and Ethics, "Final and Complete Election Results for September 15, 1998 Primary."

72. Hamil R. Harris and Yolanda Woodlee, "Council Could Become Majority White; Primary Results Create Possibility, Riling Barry and Causing Debate," *Washington Post,* September 17, 1998.

73. Jonetta Rose Barras, "Marion Barry Plays the Race Card Again," *Washington Times,* September 25, 1998.

74. Jonathan Tilove, "White-Majority D.C. Council Would Reflect New Politics," *Washington Post,* September 30, 1998.

75. Vanessa Williams, "Election Lacks Spark, Voter Enthusiasm; Only 39% of D.C. Electorate Turns Out," *Washington Post,* November 5, 1998.

76. David Montgomery, "D.C.'s New-Look Council Pledges an Activist Agenda," *Washington Post,* November 5, 1998.

77. Clay Risen, "Off the Mall: Throwing Off the Reins of Federal Control, the City of Washington, D.C., Makes Grand Plans for Its Neglected Neighborhoods," *Metropolis* 23, no. 11 (July 2004): 62, 65. The growing concern for pragmatic solutions to local problems evident among District of Columbia officials and politicians is beginning to be reflected in the attitudes and actions of congressional oversight committees as well. In February 2005, the House of Representatives abolished a separate Subcommittee on District Affairs of its Appropriation Committee for the first time since 1886. This move folds D.C. affairs into the jurisdiction of the appropriations subcommittee responsible for transportation, treasury, housing, and the judiciary. This change limits House oversight of the D.C. government. The move was seen generally as a reduction in congressional control over D.C. affairs. The Senate maintained its previous committee structure, including a panel with responsibility for the District of Columbia. Spencer S. Hsu,

"City Loses Its Panel in House Committee; Officials Expect Less Interference," *Washington Post,* February 16, 2005.

78. "Editorial: DC: WASA's Water Woes," *Washington Post,* February 4, 2004; Lyndsey Layton and Manny Fernandez, "Street Closing Irks D.C. Leaders," *Washington Post,* August 3, 2004; and "Editorial: 'Literally Falling Apart,'" *Washington Post,* December 26, 2004.

79. Lori Montgomery and Yolanda Woodlee, "Williams Traveled 190 Days in 2 Years: Mayor Says Trips Boost D.C. Image," *Washington Post,* November 24, 2004.

80. Lori Montgomery, "One Guarantee Sparked Larger Baseball Battle: $19 Million Penalty Field Resentment on D.C. Council," *Washington Post,* December 19, 2004.

81. "Washington Senators, aka Nationals, Nats," *BaseballLibrary.com,* available at http://www.baseball library.com/baseballlibrary/ballplayers/S/Senators_Washington.stm, 12/28/2004; "Washington Senators," *Wikipedia,* available at http://en.wikipedia.org/ wiki/Washington_ Senators, 12/28/2004.

82. Brad Snyder, *Beyond the Shadow of the Senators* (New York: McGraw-Hill, 2003).

83. Snyder, *Beyond the Shadow.*

84. "Washington Senators," *BaseballLibrary.com.*

85. "Washington Senators," *BaseballLibrary.com;* Snyder, *Beyond the Shadow.*

86. "Washington Senators," *Wikipedia.*

87. "Montreal Expos,"*BaseballLibrary.com,* http://www.baseballlibrary.com/base balllibrary/ballplayers/M/Montreal_Expos.stm, 12/29/2004; "Washington Nationals," *Wikipedia,* http://en.wikipedia.org/wiki/Montreal_Expos, 12/29/2004.

88. "Washington Nationals," *Wikipedia.*

89. "Washington Nationals," *Wikipedia.*

90. "Washington Nationals," *Wikipedia.*

91. "Washington Nationals," *Wikipedia.*

92. "Washington Nationals," *Wikipedia.*

93. "MCI Center," *Arenas by Munsey & Suppes* Web site, http://basketball.ballparks .com/NBA/misc/presentEASTmenu.thm, 12/29/2004.

94. "FedEx Field," *Arenas by Munsey & Suppes* Web site, http://football.ballparks .com/NFL/misc/presentNFCmenu.htm, 12/29/2004.

95. Lori Montgomery and Steven Goff, "New Soccer Stadium Is Also Eyed for SE," *Washington Post,* September 24, 2004.

96. Peter Whoriskey, "Ballparks in West Offer D.C. Divergent Lessons," *Washington Post,* November 15, 2004.

97. This argument was made eloquently by Washington Nationals' supporters on the D.C. Council led by Jack Evans, the Ward Two Council member, and other proponents of Williams's stadium plans, such as the distinguished *Washington Post* editor and columnist David Broder. See David S. Broder, "Take Them Out to the Ballgame," *Washington Post,* December 22, 2004.

98. Sally Jenkins, "Baseball's Mind Game,"*Washington Post,* December 21, 2004.

99. Marion Barry, "D.C. Baseball's Hidden Costs," *Washington Post,* December 14, 2004.

100. Jenkins, "Baseball's Mind Game."

101. Montgomery, "One Guarantee Sparked Larger Baseball Battle."

102. Eric Fisher, "Ballpark Approval Coming Today," *Washington Times,* December 14, 2004.

103. Fisher, "Ballpark Approval Coming Today."

104. Barry, "D.C. Baseball's Hidden Costs."

105. "Baseball in D.C., Operation Shuts Down (At Least for Now)," *SBN Sports Business News* Web site, http://www.sportsbusinessnews.com/index.asp?story_id=42291.

106. David Nakamura and Thomas Heath, "Amended Deal on Stadium Approved: Council Seals Return of Baseball to D.C.," *Washington Post,* December 22, 2004.

107. Associated Press, "D.C. Mayor Signs Legislation for Baseball," December 29, 2004; available at http://abcnews.go.com/Sports/wireStory?id=369129.

108. In mid-February, the Nationals opened spring training in Florida, playing their final preseason exhibition game at Robert F. Kennedy Memorial Stadium in Washington on April 2. The team lost the opening game of the season in Philadelphia on April 4, before returning home to defeat the Arizona Diamondbacks by a score of 5–3 in the first major league baseball game played in Washington in nearly three and a half decades. David Sheinin, "Emerging from the Shadows: Under the Florida Sun, Nationals Open First Training Camp," *Washington Post,* February 16, 2005; Berry Svrluga, "Washington Springs Forward to Baseball, but Lineup Questions Linger as Nats Get a Final Tuneup," *Washington Post,* April 4, 2005; Svrluga, "At Long Last, Nats Lose: On Opening Day, Washington's First Baseball Team in 34 Years Slips on the Road," *Washington Post,* April 5, 2005; and David Nakamura, "Washington Cheers as Long-Sought Team Makes Winning Debut," *Washington Post,* April 15, 2005.

109. Thomas Boswell, "A Big Reality Check," *Washington Post,* December 22, 2004.

110. Boswell, "Big Reality Check."

111. Marc Fisher, "Cropp's Tactics Saved Little Beyond Face," *Washington Post,* December 23, 2004.

112. See http://www.ihatelindacropp.com and http://lindacroppsux.blogspot.com.

113. For one concerned reaction to the bashing of Council Chair Cropp, see Colbert I. King, "Stepping Up to the Plate for the City," *Washington Post,* December 18, 2004.

114. Jenkins, "Baseball's Mind Game."

115. Lori Montgomery, "Still Tallying Hits and Errors," *Washington Post,* January 6, 2005.

116. David Nakamura, "Private Financing Deals Offered for Stadium," *Washington Post,* January 19, 2005; Nakamura, "Stadium Financing Deals Cleared: D.C. Official Approves 2 Proposals to Limit Public Commitment, Taxes," *Washington Post,* March 15, 2005; Nakamura, "Plan Could Cut D.C.'s Stadium Burden," *Washington Post,* April 12, 2005; David Nakamura, "Cropp Defined by Ballpark Push. Question for Private Funding Ends But Could Be an Issue in a Mayoral Race," *Washington Post* (June 19, 2005).

117. Barras, "Marion Barry Plays the Race Card Again."

118. Jose Suerio, "Hispanics Want a Say," *Washington Post,* November 7, 1999.

119. Sylvia Moreno, "Advocates for Immigrants Endorse D.C. Language Bill," *Washington Post,* April 9, 2003.

120. Mary Beth Sheridan, "An Agency Missing in Action: Disarray Keeps Latino Office From Making Links with the Hispanic Community—Latino Office Struggles to Reach Out; Staff Losses, Lack of Clear Direction Hobble City Agency," *Washington Post,* September 4, 2003.

121. Jennifer Frey, "Language of the Heart; for 30 Years, Sonia Gutierrez Has Helped D.C.'s Immigrants Learn in a New Land," *Washington Post,* July 6, 2002. Pamela Constable, "Newly Arrived Vietnamese Learn to Adapt; Youth Project Helps Immigrant Students Cope with Often-Inhospitable D.C. Schools," *Washington Post,* February 9, 1997.

122. Jennifer Lenhart, "Rx for Language Gap at the Doctor's Office Series: New Voices," *Washington Post,* April 27, 2000.

123. Bill B15-0139, "Language Access Act of 2003," introduced February 4, 2003; available at http://dccouncil.washington.dc.us.lims/billrecord.asp?strlegno=B15-0139.

124. These six council members were Graham, Allen, Brazil, Fenty, Mendelson, and Schwartz.

125. Moreno, "Advocates for Immigrants Endorse D.C. Language Bill."

126. Arthur Santana, "D.C. Council mandates Translators. Agencies Will Assist Non-English Speakers," *Washington Post,* April 7, 2004.

127. Mary Beth Sheridan, "D.C. Official Helps People Understand One Another: New Law Requires Agencies to Translate and Interpret," *Washington Post,* December 19, 2004.

128. Sylvia Moreno, "Residency Bill Raises Concerns about Fairness; Opponents Argue Some Students Would be Wrongly Shut Out," *Washington Post,* May 29, 2003.

129. Moreno, "Residency Bill Raises Concerns." These are, for the most part, children whose parents live in the suburbs and work in Washington or who are being cared for by their grandparents who live in town.

130. *Washington Post* staff writer, "Non-Citizens Should Get Vote, Too, Mayor Says," *Washington Post,* October 1, 2002.

131. Student Voices staff, "New Poll and Speak Out! Should Non-Citizens Have the Right to Vote?" *Washington DC Student Voices,* November 21, 2002; available at http://student-voices.org/news/index.php3?NewsID=3864.

132. Paul M. Weyrich, "Making Non-Citizens Voters: The Latest Scheme from Washington," *Free Congress Foundation* Web site, http://www.freecongress.org/commentaries/021008PW.asp.

133. Spencer S. Hsu, "House Approves District's Budget, $560 Million in Aid: Leaders Reject Letting Noncitizens Vote," *Washington Post,* July 21, 2004.

134. "Anti-Terrorism Measures and the Security of Latinos and Immigrants in DC," *D.C. North,* February 2003.

135. Brian DeBose, "Police Won't Ask Aliens of Status; Ramsey Assures City's Hispanics," *Washington Times,* July 29, 2003.

136. "Chief Offers Assurance to Immigrants," *Washington Post,* July 26, 2003.

137. "Officer Reminder: Inquiries about Immigration Status Prohibited," *Metropolitan Police Department Dispatch,* July 29, 2003; from Circular #CIR-03-12/General Order 201.26.

138. Richard Stren and Mario Polese, "Understanding the New Sociocultural Dynamics of Cities: Comparative Urban Policy in a Global Context," in *The Social Sustainability of Cities: Diversity and the Management of Change,* ed. Richard Stren and Mario Polese (Toronto: University of Toronto Press, 2000), 3–38; the quotation here is on 3.

139. Brian Ray, "Challenges to Social Inclusion in a New Immigrant Destination: Metropolitan Washington," paper presented at Woodrow Wilson Center Comparative Urban Studies Project Seminar, "Space and Identity: Concepts of Immigration and Integration in Urban Areas," Washington, January 27–28, 2005.

140. Courtland Milloy, "Jazz Provides Sisters with Key to Success," *Washington Post,* May 27, 1998; Washington Area Music Association, *WAMA News,* Fall 1998.

141. This account of the achievements of Kelly and Maze Tsefaye has appeared previously in Blair Ruble, "From 'Chocolate City, Vanilla Suburbs' to 'Rocky Road,'" *Washington Afro-American,* October 16–22, 2004.

## Chapter 6—From Red to Orange: Kyiv's Post-Soviet Municipal Regime

1. Andrey Kurkov, *Death and the Penguin* (London: Havrill Press, 1996), 168; first published in Russian as *Smert' postoronnego* (Kyiv: Alterpress, 1996).

2. This account is based on interviews and conversations with several eyewitnesses to these events.

3. This discussion is based on conversations with Oleksandr Mosijyk, Kyiv, September 8, 2003. Additional material on these events may be found in Ivan Saliy, *Oblichchia stolitsy v doliakh ii keriunykiv* (Kyiv: Mizhnarodna Fundatsiia, 2003), 149–69.

4. This account is based on a conversation with Petro Antonovich Kopil', Kyiv, January 19, 2004. Kopil', an electrical engineer specializing in microprocessing, served as deputy chair of the Kyiv City Council Executive Committee while Mosijyk served as chair of that body. Kopil', who created a sensation with his revelations about inequities in the city's late-Soviet-era housing policies, eventually went on to serve as the deputy chair of Kyiv's Moskovskii District. He cooperated closely for a while with the eventual Kyiv mayor Oleksandr Omelchenko before moving to the Southern Railway District administration. After participating in the management of extensive renovations of Kyiv's Central Railway Station, Kopil' returned to the Ukrainian National Academy of Sciences, where he became the director of Ukraine's primary research institute on microprocessing.

5. Iurii Shcherbak, *Chernobyl: A Documentary Story* (Edmonton: Canadian Institute of Ukrainian Studies Press, 1989).

6. Conversation with Kopil', Kyiv, January 19, 2004.

7. For insightful accounts of the dysfunctionality of Soviet-era city councils, see G. V. Barabashev, ed., *Rol' mestnykh sovetov v ekonomicheskom i sotsial'nom razvitii gorodov* (Moscow: Izdatel'stvo MGU, 1983); and Jeffrey W. Hahn, "Studying the Russian Experience: Lessons for Legislative Studies (and for Russia)," in *Democratization in Russia: The Development of Legislative Institutions,* ed. Jeffrey W. Hahn (Armonk, N.Y.: M. E. Sharpe, 1996), 241–61.

8. Saliy, *Oblichchia stolitsy v doliakh ii keriunykiv,* 151–52; and conversation with Kopil', January 19, 2004.

9. Conversation with Kopil', January 19, 2004.

10. Photographs of this event may be found in Saliy, *Oblichchia stolitsy v doliakh ii keriunykiv,* 168.

11. Saliy, *Oblichchia stolitsy v doliakh ii keriunykiv,* 182–84.

12. Saliy, *Oblichchia stolitsy v doliakh ii keriunykiv,* 184–88.

13. Saliy, *Oblichchia stolitsy v doliakh ii keriunykiv,* 189–203.

14. Saliy, *Oblichchia stolitsy v doliakh ii keriunykiv,* 204–17.

15. Saliy, *Oblichchia stolitsy v doliakh ii keriunykiv,* 218–54.

16. Conversation with Kopil', January 19, 2004.

17. Larysa Ivshyna, Iryna Havrylova, Liudmyla Humeniuk, Dmytro Skriabin, Kostiantyn Ryliov, and Volodymyr Zolotariov, "Ivan Saliy: I Am Now a Wise and Cautious Person," *The Day (Den'),* no. 44 (1998); available at http://day.kiev.ua/DIGEST/1998/44/economy/eco-1.htm.

18. Saliy, *Oblichchia stolitsy v doliakh ii keriunykiv,* 255–81.

19. This account is based on conversations with Oleksandr Mosijyk, Kyiv, September 8, 2003.

20. "Ukrainian Update," *RFE/RK Friends & Partners,* no. 156, part 1, 13, August

1996; available at http://www.friends-partners.org/friends/news/omri/1996/08/960813II .html.

21. Conversation with Kopil', January 19, 2004.

22. Conversations with Mosijyk, September 8, 2003; and Kopil', January 19, 2004.

23. This account is based on conversations with Oleksandr Vazylenko, vice editor of the Kyiv city newspaper *Stolytsa,* Kyiv, September 8, 2003.

24. Iryna Havrylova, "Contesting a Score of 76 to 16?," *The Day (Den'),* no. 22 (1999); available at http://day.kiev.ua/DIGEST/1999/22/titlt/tit-6.htm.

25. This account is based on conversations with Ivan Saliy, deputy mayor of the city of Kyiv, Kyiv, September 13, 2000.

26. Larysa Ivshina, Klara Gudzyk, Oleh Ivantsov, Natalia Ligachova, Serhiy Makhun, Oleksandr Fandeyev, and Iryna Chermerys, "Mayor Oleksandr Omelchenko: In Five Years We Will Be in Europe," *The Day (Den'),* no. 18 (2000); available at http://www.day.kiev.ua/DIGEST/ 2000/018/den-pln/dp1.htm.

27. Saliy, *Oblichchia stolitsy v doliakh ii keriunykiv,* 284–318.

28. Ivshyna et al., "Ivan Saliy."

29. This discussion is based on Blair A. Ruble, *Leningrad: Shaping a Soviet City* (Berkeley: University of California Press, 1990), 193–220.

30. This is based on a conversation with Oleksandr Nikolaevich Statiuk, deputy chair, Solomenskii District Council, Kyiv, January 22, 2004. In 2002, the Solomenskii District collected 500 million hryvna (about $100 million) in taxes, and received 80 million hryvna (about $16 million) in return. In 2003, the figures rose to 700 million hryvna ($140 million) being collected, with 100 million hryvna ($20 million) being returned. District officials project tax revenues of 1 billion hryvna in 2004 (just under $200 million), with 120 million hryvna ($24 million) being returned from higher authorities.

31. E.g., individual tax payers contribute 40 percent of all taxes collected in Kyiv's Solomenskii District, with the remaining 60 percent coming from land taxes, user fees, and enterprise taxes as well as other revenues (in order of descending importance). Conversation with Statiuk, January 22, 2004.

32. Conversation with Statiuk, January 22, 2004.

33. A point emphasized by Stepan Oleksandrovich Kleban, vice director, Association of Ukrainian Cities, in a conversation in Kyiv, January 23, 2004.

34. A point emphasized by Kleban, January 23, 2004.

35. This discussion is based on the above-cited conversations Mosijyk, September 8, 2003; Vazylenko, September 8, 2003; and Saliy, September 13, 2000. Further information on the legal status of the City of Kyiv may be found on the city administration's official website: www.kmv.gov.ua.

36. "Electoral Harvest Time in Kyiv," *UCIPR Research Update,* no. 8/256, February 25, 2002; available at http://old.ucipr.kiev.ua/english/rupdate/2002/02/25022002 .html.

37. "Electoral Harvest Time in Kyiv."

38. "Kuchma Reappoints Omelchenko Kyiv Mayor," *Forum,* February 19, 2002; available at http://eng.for-ua.com/news/2002/02/19/110715.html.

39. Interfax-Ukraine, "Khoroshkovskyi Enters into Kyiv Picture to Replace Omelchenko?" August 29, 2003; available at http://www2.pravda.com.ua.en/archive/ 2003/august29 /news/1.shtml.

40. Conversation with Vazylenko, September 8, 2003. The final December 30, 2003, ruling on this matter by the Constitutional Court of Ukraine upheld Mayor Omel-

chenko's right to remain in office. UNIAN News Agency, Kyiv, "Ukrainian Court Rules Kiev Mayor Can Stay On Despite Age Limit," as reported by BBC Monitoring Service in English, December 30, 2003.

41. Roman Zakaluzhny, "Recall Attempt," *Kyiv Post,* October 2, 2003.

42. Andriy Lavryk, "It's Difficult Being Mayor: Battles for the Seats of City Chairmen Are Becoming the Order of the Day," *Halytski Kontrakty,* October 6, 2003.

43. Such concern for fiscal discipline does not appear to extend to grand projects of public display so personally profitable to senior city officials, such as the construction of underground shopping malls beneath Kyiv's primary avenue, the Kreshchatyk.

44. Kennan Institute Survey of Migrant Families, Kyiv, June–November 2001, questions 3, 25, 45–65, 261. For more information about this survey, see the appendix to the present volume.

45. Kennan Institute Survey of Migrant Families, questions 157, 158, 254, 261. O. Braichevs'ka, G. Volosiuk, O. Malinovs'ka, Ia. Pilins'kii, N. Popson, and B. Rubl, *"Netraditsiyni" immigranty u Kyivy* (Kyiv: Institut Kennana / Kyiv'skii Proekt, 2003), 291–95.

46. Kennan Institute Survey of Migrant Families, questions 217, 261. Braichevs'ka et al., *"Netraditsiyni" immigranty u Kyivy,* 311–14.

47. Kennan Institute Survey of Migrant Families, questions 246, 246, 261.

48. Kennan Institute Survey of Migrant Families, questions 220, 261.

49. For further discussion of the special difficulties encountered by African residents of Kyiv, see Braichevs'ka et al., *"Netraditsiyni" immigranty u Kyivy,* 304–7. Racial hostility is a major factor in the negative perception held by African migrants of Ukraine and Kyiv. In response to a series of questions concerning the acceptance of African and Asian neighbors, nearly 60 percent of the respondents to the general survey of Kyiv residents undertaken in connection with the Kyiv Migrant Project expressed a desire not to have Africans as neighbors, two-thirds rejected Africans as potential close friends, and 80 percent did not want African migrants as family members. See Braichevs'ka et al., *"Netraditsiyni" immigranty u Kyivy,* 360–63.

50. Braichevs'ka et al., *"Netraditsiyni" immigranty u Kyivy,* 113–219.

51. Braichevs'ka et al., *"Netraditsiyni" immigranty u Kyivy,* 299–301. A point underscored in interviews throughout 2000 with those working with migrant communities in official, international, and nongovernmental agencies and offices, such as John Steven Cook, chief of mission, International Office for Migration Mission in Ukraine (interview, Kyiv, January 25, 2000); Vasyl' Stepanovych Gazhaman, former official of the Kyiv City Department of Refugee Status and Migration (interview, Kyiv, January 26, 2000); Oleksandr Piskun, editor, *Migration Issues* (interview, Kyiv, January 26, 2000); Yurii Vasyliovych Buznicki, director of the nongovernmental Center for Investigations of Migration Problems (interview, Kyiv, June 29, 2000); Mokhammed Zekriia Khamnava, Sudkhan Dzhamat, and Abdul Mokhammed Iadchari, officials of the Ukrainian Afghan Association (interview, Kyiv, June 30, 2000); Mikola Oleksiiovich Pilipchiuk, chief, Administration for Passport, Registration, and Migration Work, Ukrainian Ministry of Internal Affairs (interview, Kyiv, June 30, 2000); Rhan Viet Hung and Fam Van Bang, respectively president and board member of the Vietnamese Community Association (interview, Kyiv, September 13, 2000); and Timur D. Mamoyan, president, Ukrainian Kurdish Association (interview, Kyiv, September 15, 2000).

52. This observation is based on remarks addressed to the author on several occasions in Kyiv's City Hall.

53. Volodymyr Olexandrovitch Novyk, head, Department for Refugees and Minorities, Kyiv City State Administration (interviews, Kyiv, June 29, 2000 and May 24, 2001).

54. Braichevs'ka et al., *"Netraditsiyni" immigranty u Kyivy,* 302–10.

55. "Kandidati na post Prezidenta Ukraini," Web site of the Central Election Commission of Ukraine, http://www.cvk.gov.ua/wp003pt001f01=500rej=1.

56. Mykola Riabchuk, "The Ukrainian Fault-Line: Citizens versus Subjects," *Berliner Zeitung,* December 3, 2004; as translated by *Ukraine List* (UKL), UKL 297, December 2, 2004.

57. Riabchuk, "Ukrainian Fault-Line."

58. Elisabeth Rosenthal, *International Herald Tribune,* December 4, 2004; Nick Paton Walsh, "Fateful Dinner Party That Brought Disfigurement in its Wake," *Guardian* (U.K.), December 13, 2004; C. J. Chivers, "A Dinner in Ukraine Made for Agatha Christie," *New York Times,* December 20, 2004.

59. Francesca Mereu, "Spin Doctors Blame Yanukovich," *Moscow Times,* November 30, 2004; "Two Victors, Only One Victor," *Economist,* November 19, 2004.

60. For just a few of the particularly concise summary articles among the scores published at the time about the Russian dimension of the Ukrainian "crisis," see Alan Cullison and Guy Chazan, "Putin's Role in Ukraine Election Stirs Backlash," *Wall Street Journal,* November 29, 2004; Nikolai Petrov, "A Major Setback for Putin," *Moscow Times,* November 29, 2004; Steven Lee Myers, "Putin Scoffs at Calls for New Runoff in Ukraine," *New York Times,* December 2, 2004; Marie Mendras, "Putin's Model Fails in Ukraine," *Le Monde* (Paris), December 2, 2004, as translated by George Zakhem for *Urkaine List,* UKL 299, December 2, 2004; and Michael McFaul, "Putin Gambles Big—and Loses," *Weekly Standard,* December 13, 2004.

61. "Two Victors, Only One Victor."

62. Dominique Arel, "Breaking News," *Ukraine List,* UKL 269, November 21, 2004.

63. Arel, "Breaking News."

64. Office for Democratic Institutions and Human Rights, Organization for Security and Cooperation in Europe, press release, November 22, 2004, as reported in *Ukraine List,* UKL 269e, November 21, 2004.

65. C. J. Chivers, "Winner Is Declared in Ukraine, and Opposition Declares Fraud," *New York Times,* November 22, 2004.

66. "Tents on the Square," *Ukrains'ka pravda,* November 22, 2004; as translated in *Ukraine List,* UKL 270, November 22, 2004.

67. "The L'viv and Ivano-Frankivs'k City Councils Recognize Yushchenko as President," *Ukrains'ka pravda,* November 22, 2004; as translated in *Ukraine List,* UKL 270, November 22, 2004. Also Stefan Wagstyl and Tom Warner, "Vote Row Spreads Beyond Ukraine," *Financial Times,* November 23, 2004.

68. For a sampling of commentary on official and unofficial Russian behavior during this period, see Helen Womack, "One Election, Two Victors," *Christian Science Monitor,* November 23, 2004; Igor Torbakov, "Kremlin Recognizes Yanukovych as Ukraine's New President," *Eurasian Daily Monitor,* November 23, 2004; and "What Role Did Russia Play in Ukrainian Election Campaign?" *Izvestiia,* November 23, 2004.

69. For a sampling of contemporaneous reports of events before the critical Ukrainian Supreme Court hearing, see David Holley, "Ukraine Slips into Election Crisis," *Los*

*Angeles Times,* 22 November 2004; Sarah Rainsford, "Night of Chaos in Kyiv," BBC News, November 23, 2004; C. J. Chivers, "Ukraine's High Court Halts Move to Make Election Official," *New York Times,* November 25, 2004; Ruslan Tracz, "Kyiv Filled with Hope, Uncertainty," *Winnipeg Free Press,* November 25, 2004; Mark MacKinnon, "Ukraine's Revolt Gains Steam," *Globe and Mail,* November 26, 2004; Ivan Vasyunnyk, "The Rise of a Nation," *Wall Street Journal Europe,* November 26, 2004; "Pro-Russian Eastern Ukraine Threatens to Secede if Yushchenko Wins," *MosNews.com,* November 26, 2004; Tom Warner, "Crowds Take On a Dynamic of Their Own," *Financial Times,* November 26, 2004; Associated Press, "Ukraine's Parliament Calls Presidential Election Invalid," November 27, 2004; and Steven Lee Myers, "Parliament Says Votes in Ukraine Were Not Valid," *New York Times,* November 28, 2004.

70. For an overview of these events, see such reports as C. J. Chivers, "Supporters of President-Elect in Ukraine Push Back," *New York Times,* November 29, 2004; Steve Lee Myers, "Departing Ukrainian President Would Support New Election," *New York Times,* November 29, 2004; Chivers, "Ukraine's Intramural Contest: Entrenched Interests Scramble," *New York Times,* December 1, 2004; Andrew Osborn, "MPs Speak, the Candidates Agree, and Ukraine Hopes for an End to the Crisis," *Independent* (U.K.), December 2, 2004; Myers, "Putin Scoffs at Calls for New Runoff in Ukraine," *New York Times,* December 2, 2004; and Kim Murphy, "Eastern Ukraine Is United Behind Yanukovych," *Los Angeles Times,* December 3, 2004.

71. "The Decision of the Supreme Court of Ukraine with Respect to the Allegations of Election Violations on November 21, 2004, Decision rendered on December 3, 2004," as translated by Nikolai Bilaniuk in *Ukraine List,* UKL 298, December 6, 2004.

72. "Decision of the Supreme Court of Ukraine."

73. "Decision of the Supreme Court of Ukraine."

74. "Decision of the Supreme Court of Ukraine."

75. For a sampling of reports during this period, see, C. J. Chivers, "It Was Dec. 3, but in Kyiv, New Year Began Yesterday," *New York Times,* December 4, 2004; and Stefan Wagstyl and Tom Warner, "Dancing in the Streets as 'a Nation Is Born,'" *Financial Times,* December 4, 2004.

76. Steven Lee Myers, "Parliament in Kiev Begins Re-Examining Flawed Election Laws," *New York Times,* December 5, 2004; Nadia Diuk, "In Ukraine, Homegrown Freedom," *Washington Post,* December 4, 2004; Myers, "Ukraine Leader Advises Boycott of Runoff Vote," *New York Times,* December 6, 2004; Gwendolyn Sasse, "There Are Hurdles Aplenty in Ukraine's Presidential Marathon but an End Is in Sight," *The Guardian* (U.K.), December 6, 2004; Myers, "No Deal Yet in Ukraine after Talks on Political Changes," *New York Times,* December 7, 2004.

77. C. J. Chivers, "Ukraine's Sharp Turn toward the West," *New York Times,* December 9, 2004.

78. For a sampling of press reports during this period, see Anne Applebaum, "Russia's Last Stand," *Washington Post,* December 15, 2004; Peter Finn, "Ukrainian Premier Foresees New Crisis," *Washington Post,* December 17, 2004; David Holley, "Yanukovich Stokes Backers' Fears," *Los Angeles Times,* December 16, 2004); Tom Parfitt, "Yanukovich 'to Twist Fresh Ukraine Ballot,'" *Sunday Telegraph* (U.K.), December 19, 2004; Yaroslav Trofimov, "Yushchenko's Formidable Challenge," *Wall Street Journal,* December 20, 2004; Tom Warner, "Ukraine Poll Contenders in Charged Debate," *Financial Times,* December 20, 2004; and Fred Weir and Howard LaFranchi,

"Who's Influencing Ukraine's Vote? As Thousands of Election Monitors Arrive for Sunday's Vote, Critics Complain of Western Interference," *Christian Science Monitor,* December 23, 2004. The text of the debate, in English translation, may be found on *Ukraine List,* UKL 322, December 23, 2004. Summaries of the final Constitutional Court Decisions altering some of the compromise election procedures passed by Parliament on December 8 may be in Steven Lee Meyers, "Ukrainian Court Overturns Homebound Voter Limit," *New York Times,* December 26, 2004.

79. Dominique Arel, "The Polarization Is Indeed More Pronounced," *Ukraine List,* UKL 327, December 27, 2004; Associated Press, "Yanukovich Refuses to Concede Ukrainian Elections," December 27, 2004; "Pisumki golosuvaniia po regionakh Ukraini: Porivniannia—Vibori Prezidenta Ukraini—Povtorne golosuvaniia 26.12.2004," Web site of the Central Election Commission of Ukraine, http://www.cvk.gov.ua/wp333 pt001f01=500rej=1.

80. Peter Finn, "West-Leaning Leader Appears Headed for Win in Ukraine Vote," *Washington Post,* December 27, 2004.

81. Steven Lee Myers, "Ukraine President Sworn In, Promising to Promote Unity," *New York Times,* January 24, 2005.

82. "Security Service of Ukraine Supports Yushchenko," *Ukrains'ka pravda,* November 25, 2004, as translated by Yulia Tarotska for *Ukraine List,* UKL 281, November 25, 2005.

83. Stefan Wagstyl, Chrystia Freeland, and Tom Warner, "Ukrainian President Spurned Pressure over Protesters," *Financial Times,* December 13, 2004.

84. C. J. Chivers, "How Top Spies in Ukraine Changed the Nation's Path," *New York Times,* January 17, 2005.

85. These remarks are based on a series of conversations with William Green Miller, who had served as the U.S. ambassador to Ukraine between 1993 and 1998. Millier, who subsequently became a senior scholar at the Woodrow Wilson International Center for Scholars, participated in many of the events described above. His thoughtful reflections on the Orange Revolution were presented in public at a Woodrow Wilson Center Director's Forum on January 6, 2005. The text of his remarks is available on the Wilson Center's Web site at http://www.wilsoncenter.org/index.cfm?fuseaction=news.print &news_id=104700&stopplay, 1/18/2005.

86. Chivers, "How Top Spies in Ukraine Changed Nation's Path."

87. "L'viv and Ivano-Frankivs'k City Councils Recognize Yushchenko as President"; and Wagstyl and Warner, "Vote Row Spreads beyond Ukraine."

88. "Tymoshenko's Cabinet and Governors," as posted on http://www2.pravda .com.ual, February 4, 2005, and translated by Olga Bogatyrenko for *Ukraine List,* UKL 338, February 8, 2005.

89. Valeriy Panyushkin, "Slava Ukrainy!" as posted on *Gazeta.ru,* November 26, 2004.

90. Neal Ascherson, "Is This To Be the Story?" *London Review of Books,* January 6, 2005.

91. "V 'oranzhevoi revoliutsii' v Ukraine uchastvovali ne tol'ko predtabiteli beloi rasy," *Korrespondent.net,* December 23, 2004, available at http://www.korrespondent .net/main/110107.

92. Braichevs'ka et al., *"Netraditsiyni" immigranty u Kyivy,* 50–56.

93. More than a quarter of Chinese migrants and a third of Vietnamese worked in

manual rural or industrial labor before coming to Ukraine. Braichevs'ka et al., *"Netra-ditsiyni" immigranty u Kyivy,* 60–65.

94. William Arutunovych Amalian, chief physician of the Nefto-Gaz Industry (interview, Kyiv, May 29, 2001).

95. Amalian, interview, May 29, 2001.

96. Braichevs'ka et al., *"Netraditsiyni" immigranty u Kyivy,* 163–65.

97. Braichevs'ka et al., *"Netraditsiyni" immigranty u Kyivy,* 164–65.

98. Braichevs'ka et al., *"Netraditsiyni" immigranty u Kyivy,* 164.

99. Braichevs'ka et al., *"Netraditsiyni" immigranty u Kyivy,* 310–11.

100. Braichevs'ka et al., *"Netraditsiyni" immigranty u Kyivy,* 292–94.

101. Nancy E. Popson and Blair A. Ruble, "Kyiv's Nontraditional Immigrants," *Post-Soviet Geography and Economics* 44, no. 5 (2000): 365–78; as well as Popson and Ruble, "A Test of Urban Social Sustainability: Societal Responses to Kyiv's 'Non-Traditional' Migrants," *Urban Anthropology* 30, no. 4 (2001): 381–409.

102. D. Trusova, "Iz devatsi deti vos'merykh v sem'e rusama uzhe vyrezali," *Fakty i kommentarii,* June 1, 2000.

103. Braichevs'ka et al., *"Netraditsiyni" immigranty u Kyivy,* 198.

104. Braichevs'ka et al., *"Netraditsiyni" immigranty u Kyivy,* 198.

105. Trusova, "Iz devatsi deti vos'merykh v sem'e rusama uzhe vyrezali."

106. Popson in Popson and Ruble, "Test of Urban Social Sustainability," 394–95.

107. Popson in Popson and Ruble, "Test of Urban Social Sustainability," 394–95.

108. Braichevs'ka et al., *"Netraditsiyni" immigranty u Kyivy,* 204–8.

109. Braichevs'ka et al., *"Netraditsiyni" immigranty u Kyivy,* 203–5.

110. Braichevs'ka et al., *"Netraditsiyni" immigranty u Kyivy,* 208.

111. The average teachers wage at the time of the Kennan Migrant Interviews was about $22.00 per month. Popson in Popson and Ruble, "Test of Urban Social Sustainability," 405.

112. Kennan Institute Survey of Migrant Families, questions 181, 261.

113. Kennan Institute Survey of Migrant Families, questions 182, 261.

114. Kennan Institute Survey of Migrant Families, questions 184, 261.

115. Braichevs'ka et al., *"Netraditsiyni" immigranty u Kyivy,* 347–49.

116. Braichevs'ka et al., *"Netraditsiyni" immigranty u Kyivy,* 301–4.

117. Fewer than 5 percent of the city's 600,000 or so annually registered crimes in recent years have been committed by transnational migrants. Braichevs'ka et al., *"Ne-traditsiyni" immigranty u Kyivy,* 352–54.

118. A number of participants in municipal politics were reporting in late 2003 that the notion of singling migrants out for attack in political campaigns has been encouraged by American consultants who have told their Kyiv clients that the issue can generate immediate attention and elicit political support.

119. E.g., Afghan leaders and local officials concur that the per capita crime rate among Kyiv's Afghan community has remained far lower than that of native-born city residents (interview, Mokhammed Zekriia Khamnava, Sudkhan Dzhamat, and Abdul Mokhammed Iadchari, Afghan community leaders, Kyiv, June 30, 2000; and interview, Yurii Vasyliovych Buznicki, chairman, Ukrainian Charitable Foundation "Migration," Kyiv, June 29, 2000).

120. M. O. Shul'ha, R. P. Shul'ha, N. L. Boiko, and T. Yu. Zagorodnyuk, *Vyvchen-nia vplyvu zovnishn'oyi mihratsii 91–1996 rr: Na sminy etnichnoho skladu naselennya*

*Ukrayiny ta ii rehioniv.* (Kyiv: International Organization for Migration, 1998), 68–69; N. A. Shul'ga, *Velikoe pereselenie: Repatrianty, bezhentsy, trudovye migranty* (Kyiv: Institut sotsiologii, NAN Ukrainy, 2002), 213–26, 605–54; Mary Frances Muzzi, "Ukrainian Press Surveys, March–May 2002," unpublished manuscript, Kennan Institute, Washington, 2001; Olena Nikolayenko, "Migration in Ukraine, 1998–2001: Annotated Bibliography," unpublished manuscript, Kennan Institute, Washington, 2001.

121. Shul'ga, *Velikoe pereselenie,* 214–17.

122. Shul'ga, *Velikoe pereselenie,* 216.

123. Shul'ga, *Velikoe pereselenie,* 217–24.

124. Muzzi, "Ukrainian Press Surveys"; Nikolayenko, "Migration in Ukraine."

125. Braichevs'ka et al., *"Netraditsiyni" immigranty u Kyivy,* 354–56.

126. Shul'ga, *Velikoe pereselenie,* 219–20.

127. Karim H. Karin, "The Multiculturalism Debate in Canadian Newspapers: The Harbinger of a Political Storm?" *Journal of International Migration and Integration / Revue de l'Integration et de la Migration Internationale* 3, nos. 3–4 (2002): 439–55.

128. Vasyl' Stepanovych Gazhaman, former official of the Kyiv City Department of Refugee Status and Migration, interview.

129. Braichevs'ka et al., *"Netraditsiyni" immigranty u Kyivy,* 355.

130. Richard Stren and Mario Polese, "Understanding the New Sociocultural Dynamics of Cities: Comparative Urban Policy in a Global Context," in *The Social Sustainability of Cities: Diversity and the Management of Change,* ed. Richard Stren and Mario Polese (Toronto: University of Toronto Press, 2000), 3–38; the quotation here is on 3.

131. This observation is based on the overall results of the Kennan Institute Survey of Migrant Families, as reported in Braichevs'ka et al., *"Netraditsiyni" immigranty u Kyivy.*

132. For further discussion of Ukraine's relatively liberal legal environment, see Oxana Shevel's perceptive analysis comparing the performance of indigenous and international institutions in dealing with migrant issues in Russia, Ukraine, and various neighboring states, argues that international organizations in general—and the United Nations High Commission for Refugees, in particular—have been unusually effective in Ukraine due to their relatively late entry into the country, by which time lessons learned from previous experiences in Russia and elsewhere enabled both international and Ukrainian officials to respond more effectively to one another. See Oxana Shevel, *International Influence in Transition Societies: The Effect of UNHCR and Other IOs on Citizenship Policies in Ukraine,* Rosemary Rogers Working Paper 7 (Cambridge, Mass.: Inter-University Committee on International Migration, 2000).

133. Polling on such volatile issues as social tolerance either did not take place—or the results were not published—prior to the late *perestroika* period of the Gorbachev era. Initial polling results on tolerance are summarized in review articles from the period, such as Natalya Panina, "National Tolerance and Relations between Nationalities in Ukraine," *Political Portrait of Ukraine Bulletin,* no. 5 (1995: 47–68.

134. Panina, "National Tolerance."

135. Braichevs'ka et al., *"Netraditsiyni" immigranty u Kyivy,* 360–61.

136. Braichevs'ka et al., *"Netraditsiyni" immigranty u Kyivy,* 360–61.

137. Braichevs'ka et al., *"Netraditsiyni" immigranty u Kyivy,* 359–66.

138. Braichevs'ka et al., *"Netraditsiyni" immigranty u Kyivy,* 364.

139. For further discussion of this point see Viktor Stepanenko, "Identities and Lan-

guage Politics in Ukraine: The Challenges of Nation-State Building," in *Nation-Building, Ethnicity and Language Politics in Transition Countries,* ed. Fariman Daftary and François Grin (Budapest: Open Society Institute Local Government and Public Service Reform Initiative / European Centre for Minority Issues, 2003), 109–35.

140. Braichevs'ka et al., *"Netraditsiyni" immigranty u Kyivy,* 365–66.

141. Braichevs'ka et al., *"Netraditsiyni" immigranty u Kyivy,* 364–65.

142. Braichevs'ka et al., *"Netraditsiyni" immigranty u Kyivy,* 359.

143. Migrant Profile, Kennan Institute Survey of Migrant Families.

144. As reported in personal communication by Nikolai Shul'ga, deputy director of the Institute of Sociology of the National Academy of Sciences of Ukraine, Kyiv, January 21, 2004. This figure is somewhat misleading because it does not include native-born Ukrainians who departed the country unofficially.

# Chapter 7—Diversity Capital Created

1. Jorge Luis Borges, "Unworthy," in Jorge Luis Borges, *Collected Fiction* (New York: Penguin, 1998), 352.

2. The notion of a forced embrace of diversity as a consequence of elite divisions and social fragmentation is essential to, e.g., Thomas Bender's portrayal of New York / New Amsterdam. See Thomas Bender, *The Unfinished City: New York and the Metropolitan Idea* (New York: New Press, 2002), 192.

3. For further discussion of the concept of "pragmatic pluralism," see Blair A. Ruble, *Second Metropolis: Pragmatic Pluralism in Gilded Age Chicago, Silver Age Moscow, and Meiji Osaka* (New York and Washington: Cambridge University Press and Woodrow Wilson Center Press, 2001; Baltimore and Washington: Johns Hopkins University Press and Woodrow Wilson Center Press, 2004).

4. Richard Stren and Mario Polese, "Understanding the New Sociocultural Dynamics of Cities: Comparative Urban Policy in a Global Context," in *The Social Sustainability of Cities: Diversity and the Management of Change,* ed. Richard Stren and Mario Polese (Toronto: University of Toronto Press, 2000), 3–38; the quotation here is on 3.

5. Myron Weiner, *The Global Migration Crisis: Challenge to States and to Human Rights* (New York: HarperCollins, 1995), x.

6. William Rogers Brubaker, *Immigration and the Politics of Citizenship in Europe and North America* (New York: University Press of America and German Marshall Fund of the United States, 1989), 1.

7. Nancy Foner, ed., *New Immigrants in New York* (New York: Columbia University Press, 1987), 17–29; and Foner, *From Ellis Island to JFK: New York's Two Great Waves of Immigration* (New Haven, Conn.: Yale University Press and Russell Sage Foundation, 2000).

8. Larry S. Bourne, "Designing a Metropolitan Region: The Lessons and Lost Opportunities of the Toronto Experience," in *The Challenge of Urban Government: Policies and Practices,* ed. Mila Freire and Richard Stren (Washington, D.C.: World Bank, 2001), 27–46; the quotations here and just below are on 27.

9. Foner, *From Ellis Island to JFK;* Thomas Sowell, *Migrations and Cultures: A World View* (New York: Basic Books, 1996), 25–32.

10. Elke Laur, "Perceptions linguistiques à Montréal, tome 1 de 2," Ph.D. dissertation in anthropology, University of Montreal, Montreal, 2001, 2.

11. Kengo Akizuki, "Immigrant Communities and Governance in Japan," paper prepared for the "Migration and City" Seminar sponsored by the Suntory Foundation, Graduate School of Law, Kyoto University, Kyoto, May 21, 2004, 7–9.

12. As cited in Akizuki, "Immigrant Communities," 4–5.

# Index

255